Paediatric Nephrology: A Research Update

Contributions to Nephrology

Vol. 67

KARGER

Basel · München · Paris · London · New York · New Delhi · Singapore · Tokyo · Sydney

21st Annual Meeting of the European Society for Paediatric Nephrology, Budapest, September 2–6, 1987

Paediatric Nephrology: A Research Update

Volume Editors
Domokos Boda, Sándor Túri, Szeged

54 figures and 51 tables, 1988

KARGER

Basel · München · Paris · London · New York · New Delhi · Singapore · Tokyo · Sydney

Contributions to Nephrology

Library of Congress Cataloging-in-Publication Data
European Society for Paediatric Nephrology. Meeting (21st: 1987: Budapest, Hungary)
Paediatric nephrology: a research update / 21st Annual Meeting of the European Society for Paediatric Nephrology, Budapest, September 2–6, 1987; volume editors, Domokos Boda, Sándor Túri.
(Contributions to nephrology; vol. 67)
Includes bibliographies and index.
1. Pediatric nephrology – Congresses. I. Boda, Domokos. II. Túri, Sándor, 1948–.
III. Title. IV. Series. V. Series: Contributions to nephrology; v. 67.
[DNLM: 1. Kidney Diseases – in infancy & childhood – congresses.
W1 CO778UN v. 67 / WS 320 E89p 1987]
RJ476.K5E87 1987 618.92′61 – dc19
ISBN 3-8055-4689-0

Bibliographic Indices
This publication is listed in bibliographic services, including Current Contents®
and Index Medicus.

Drug Dosage

The authors and the publisher have exerted every effort to ensure that drug selection and dosage set forth in this text are in accord with current recommendations and practice at the time of publication. However, in view of ongoing research, changes in government regulations, and the constant flow of information relating to drug therapy and drug reactions, the reader is urged to check the package insert for each drug for any change in indications and dosage and for added warnings and precautions. This is particularly important when the recommended agent is a new and/or infrequently employed drug.

Printed in Switzerland by Thür AG Offsetdruck, Pratteln
ISBN 3-8055-4689-0

Contents

Prenatal Diagnosis of Urological and Nephrological Diseases

Follow-Up Studies

Foreword

The lectures featured in this publication were delivered at the 21st Annual Meeting of the European Society of Paediatric Nephrology in Budapest. The organizers of this conference had one main aim: to include those topics which best represent the most recent advances and perspectives in paediatric nephrology, both from the theoretical and the practical aspects, and to select the most outstanding contributions to these topics.

A large proportion of the papers published here deal with the latest research results achieved in connection with vasoactive mediators, kidney function and renal diseases.

Investigations relating to the effects of prostanoids, other vasoactive hormones, and, particularly, natriuretic peptides, have opened up a completely new chapter on the pathophysiology of the kidney and the relevant clinical practice.

Results that are similarly of a determining nature as concerns both theory and practice were also reported in lectures dealing with the membrane functions of the cells and with the effects of immunological factors.

Especially noteworthy are papers which were presented in the field of neonatal nephrology, demonstrating that disturbances of kidney function are currently prominent problems in disturbances of the adaptation of neonates to extrauterine life.

Extremely valuable information was provided by the reports discussing the possibilities of the prenatal diagnosis of renal diseases, in particular through the use of ultrasound and other up-to-date techniques.

By virtue of their nature, the various international scientific conferences allow a many-faceted comparison of experience on topical questions, whereby far-reaching conclusions can be drawn and trends for the future can be decided. In this context, it may be pointed out that certain of the papers in this publication were a direct result of inter-institutional collaboration.

The editors are particularly grateful to those authors who have provided especially revised manuscripts for this volume of the Karger series 'Contributions to Nephrology'.

Domokos Boda
Sándor Túri

Contr. Nephrol., vol. 67, pp. 1–8 (Karger, Basel 1988)

Atrial Natriuretic Peptide and Other Vasoactive Hormones in Volume Regulation[1]

W. Rascher, M. Bald, T. Tulassay, K. Schärer
Division of Pediatric Nephrology, Universitäts-Kinderklinik, Heidelberg, FRG

Sodium balance is very well maintained in healthy subjects with renal sodium and water excretion adjusted closely to sodium and water intake, thereby preserving a constant extracellular fluid volume. The volume of this compartment is mainly determined by the quantity of sodium. Plasma volume is determined by the total extracellular fluid volume and the partitioning of this volume between extravascular and intravascular compartments according to the Starling relationship. Volume receptors detect changes in extracellular fluid volume relative to intravascular and interstitial capacitance and various renal effector mechanisms finally modify the rate of sodium excretion by the kidneys to meet the demand of volume homeostasis [14].

Various hormones with vasoactive properties play a role as renal effector mechanisms for body fluid homeostasis. In this brief communication we summarize some aspects of atrial natriuretic peptide in natriuresis induced by volume expansion and the regulation of this peptide in children with chronic renal failure. Furthermore, the role of various vasoactive hormones in edema formation of children with nephrotic syndrome is discussed.

[1] Supported by the Deutsche Forschungsgemeinschaft (Ra 326/1-2).

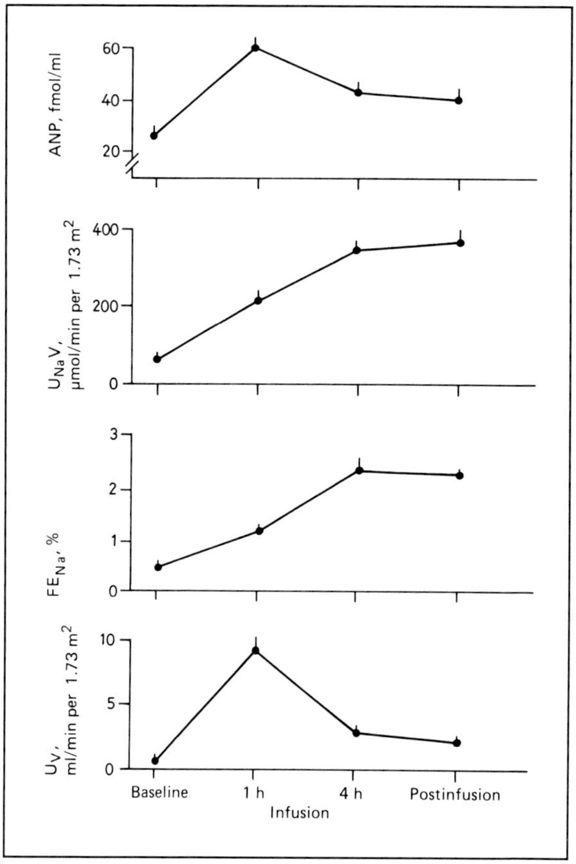

Fig. 1. Changes in plasma concentration of ANP, urinary sodium excretion ($U_{Na}V$), fractional excretion of filtered sodium (FE_{Na}), and urine flow rate (U_V) in 7 healthy volunteers following hypotonic extracellular volume expansion. Volume expansion was induced by drinking 20 ml/kg tap water and subsequent intravenous infusion of 2 liters of 0.9% saline solution over a period of 4 h.

Atrial Natriuretic Peptide and Volume Expansion

The atrial myocytes are the site of synthesis and storage of a hormone called atrial natriuretic peptide (ANP) [1, 3, 8]. It is released from cardiac atria following atrial distension [7]. Intravenous administration of human α-ANP increased urinary sodium and water excretion [12, 16].

Central volume expansion induced by saline infusion in normal volunteers [13, 18], by head-out water immersion [5] or by infusion of hyperoncotic human serum albumin solution in children with nephrotic syndrome [15], increased plasma ANP and induced natriuresis and diuresis. This indicates that expansion of the blood volume provokes the release of ANP from human atria into the circulation. ANP, in turn, promotes sodium and water excretion.

In order to study the relationship between ANP release and natriuretic response, we investigated 7 healthy, normal adult volunteers during hypotonic extracellular volume expansion (ECVE). ECVE was induced by drinking 20 ml/kg of tap water after 12 h of dehydration, followed by intravenous infusion of 2 liters of 0.9% saline over a period of 4 h. Hypotonic ECVE caused a two- to threefold increase in the plasma concentration of ANP, whereas plasma renin activity, aldosterone, vasopressin and noradrenaline levels fell. In response to ECVE, creatinine clearance rose, as did urine flow rate, urinary sodium excretion and fractional excretion of sodium (FE_{Na}) (fig. 1). There was a remarkable dissociation between the ANP response and urinary sodium excretion following ECVE. Whereas sodium excretion rose progressively over a period of 5 h, plasma ANP concentration rose threefold in the first hour of ECVE, and tended to decrease during the next 4 h. Plasma ANP appeared to be related to the diuretic response of ECVE. Thus, our results are compatible with the concept that increased ANP secretion may play a role in immediate increase in sodium excretion after hypotonic ECVE. However, they also suggest that other mechanisms may be more important for longer term natriuresis.

ANP in Children with Chronic Renal Failure

We have recently shown that children with volume overload secondary to end-stage renal disease had higher predialysis plasma concentrations of ANP than normovolemic children with advanced chronic renal failure and than normal controls [10]. Fluid removal by hemodialysis resulted in a dramatic fall of plasma ANP. This has been confirmed subsequently in adult patients [4, 6, 17].

Eight adolescents on regular intermittent hemodialysis treatment were studied during 1 h of sequential ultrafiltration followed by 3 h on hemodialysis. During removal of fluid excess by 1 h of sequential ultrafiltration plasma ANP fell from 123.8 ± 36.6 to 45.5 ± 9.3 fmol/ml ($\bar{x}$ ± SEM).

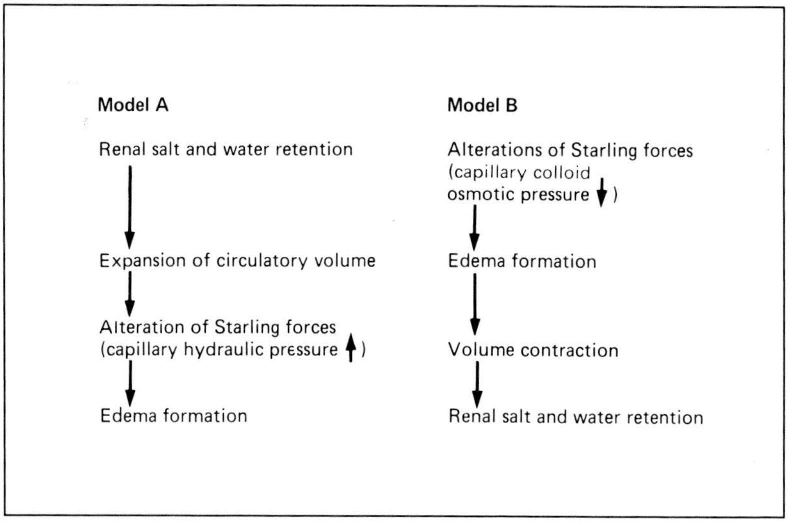

Fig. 2. Models of edema formation.

Subsequent hemodialysis combined with fluid removal for 3 h decreased ANP further down to 29.6 ± 4.6 fmol/ml. Thus, the decrease of elevated ANP during hemodialysis is mainly due to reduction of volume overload. Thus, plasma ANP appears to be a sensitive parameter for the evaluation of fluid retention in children with end-stage renal disease.

Vasoactive Hormones during Edema Formation in Children with Nephrotic Syndrome

Volume regulation is disturbed in edematous children with nephrotic syndrome. Theoretically, edema may develop by two mechanisms, i.e. when capillary hydraulic pressure increases as a result of constant elevation of plasma volume (model A) or when colloid osmotic pressure drops (model B) (fig. 2).

Neither increased capillary filtration nor decreased colloid osmotic pressure alone are sufficient to produce edema. In the presence of normal renal function a rise in capillary filtration with subsequent expansion of extracellular fluid volume results in increased natriuresis [14]. On the other hand, low plasma colloid osmotic pressure per se does not produce edema, as demonstrated by patients with congenital analbuminemia [2]. Thus, besides capillary factors, impairment of renal sodium and water excretion contributes to edema formation. According to the classical concept of salt and water retention in patients with nephrotic syndrome reduced plasma volume should stimulate various vasoactive and volume-retaining hormone systems, which in turn promote salt and water retention. We have recently studied the role of various vasoactive hormone systems in edematous children with nephrotic syndrome [9, 11, 15]. In contrast to previous investigations in adults, we studied a homogenous group of children and adolescents suffering from steroid-sensitive nephrotic syndrome, with histologically documented minimal change disease. Patients were studied during the starting phase of edema formation, defined as an increase in body weight of more than 5% within 1 week and/or manifest edema since 3 days.

Compared to 15 healthy, age-matched children (median age 12.2 years, range 6–18 years), 17 nephrotic children (median age 11.0 years, range 2–17 years) during edema formation had higher plasma concentrations of arginine-vasopressin, noradrenaline, aldosterone and plasma renin activity [9]. Basal plasma ANP did not differ between 9 healthy controls and 9 nephrotic children. Thus, various vasoactive hormones are stimulated during edema formation and may contribute to salt and water retention.

To further evaluate the mechanism of raised hormone secretion we compared children in remission of a nephrotic syndrome to children during relapse with edema formation with respect to plasma vasopressin and water metabolism. Edematous children had marked proteinuria (5.6 ± 2.7 g/24 h per 1.73 m^2, $\bar{x} \pm$ SD), reduced serum protein and albumin concentration, and elevated serum cholesterol [9]. Serum creatinine, serum urea, serum potassium and creatinine clearance did not differ. However, serum sodium, chloride, calcium and plasma osmolality were lower and hematocrit higher in edematous children compared to children in remission [9]. The mean plasma vasopressin concentration was markedly elevated in edematous children.

Thus, vasopressin concentration is elevated during edema formation in children with nephrotic syndrome when compared either to healthy normal children or to children with nephrotic syndrome in remission. Furthermore, edematous children are slightly hypoosmolar and hyponatremic, indicating a nonosmotic stimulus for vasopressin secretion. Higher hematocrit values in proteinuric nephrotic children point to hemoconcentration and a reduced circulatory blood volume as a cause for elevated vasopressin secretion in these patients. This is corroborated by the observations of increased activity of various vasoactive hormone systems stimulated by volume reduction, such as the renin-angiotensin-aldosterone system and plasma noradrenaline.

Increase in effective circulatory blood volume should be able to suppress vasopressin secretion and release of renin, aldosterone and noradrenaline into the circulation, and consequently improve salt and water excretion. Therefore, we studied in edematous nephrotic children, the hormonal and renal responses to head-out water immersion and infusion of hyperoncotic human serum albumin solution, both procedures known to increase central blood volume.

Head-out water immersion was studied in 8 edematous children with nephrotic syndrome [11]. During 3 h of water immersion the hematocrit and serum protein concentration fell, indicating central volume expansion. Mean urine flow rose four- to fivefold as did urinary sodium excretion. Urinary potassium excretion rose threefold and osmolar clearance and creatinine clearance doubled. Plasma osmolality remained stable and urine osmolality fell. Elevated plasma concentrations of vasopressin, aldosterone, noradrenaline and plasma renin activity decreased during immersion and rose after immersion was ended [11].

Volume expansion induced by infusion of hyperoncotic serum albumin solution (20% solution, 1 g/kg per 90 min) resulted in a fall of elevated plasma concentrations of vasopressin, aldosterone, noradrenaline, and plasma renin activity. Plasma osmolality did not change. The plasma ANP concentration rose fivefold in response to albumin infusion [15]. Hyperoncotic albumin induced diuresis, natriuresis and a rise in creatinine and osmolar clearances. Urine osmolality fell, but free water clearance did not change. Changes induced by hyperoncotic albumin resembled those observed during water immersion.

Both induced central volume expansion by water immersion [11] and infusion of hyperoncotic serum albumin solution [15] reduced hematocrit values and resulted in a drop of elevated plasma levels of various vasoactive hormones to near-normal values and improved salt and water excretion.

Thus, our studies point to an important role of reduced circulating blood volume with subsequent activation of vasoactive and sodium-retaining hormones for sodium and water retention and therefore for edema formation in children with nephrotic syndrome (according to model B from figure 2). This does not exclude that in patients with glomerular disease and nephrotic syndrome edema formation may be related to model A.

References

1 Ballermann, B.J.; Brenner, B.M.: Biologically active atrial peptides. J. clin. Invest. *76:* 2041–2048 (1985).

2 Bennhold, H.; Klaus, D.; Scheinlen, P.G.: Volume regulation and renal function in analbuminemia. Lancet *ii:* 1169–1170 (1960).

3 De Bold, A.J.; Borenstein, H.B.; Veress, A.T.; Sonnenberg, H.: A rapid and potent natriuretic response to intravenous injection of atrial myocardial extract in rat. Life Sci. *28:* 89–94 (1981).

4 Eisenhauer, T.; Talartschik, J.; Scheler, F.: Detection of fluid overload by plasma concentration of human atrial natriuretic peptide in patients with renal failure. Klin. Wschr. *64:* suppl. VI, pp. 68–72 (1986).

5 Epstein, M.; Loutzenhiser, R.; Friedland, E.; Aceto, R.M.; Camargo, M.J.F.; Atlas, S.A.: Relationship of increased plasma atrial natriuretic factor and renal sodium handling during immersion-induced central hypervolemia in normal humans. J. clin. Invest *79:* 738–745 (1987).

6 Hartter, E.; Pacher, R.; Frass, M.; Woloszczuk, W.; Leithner, C.: Plasma levels of atrial natriuretic peptide in volume expanded patients; response to fluid removal by continuous pump driven hemofiltration. Klin. Wschr. *64:* suppl. VI, pp. 112–114 (1986).

7 Lang, R.E.; Thölken, H.; Ganten, D.; Luft, F.C.; Ruskoaho, H.; Unger, T.: Atrial natriuretic factor. A circulating hormone stimulated by volume loading. Nature, Lond. *314:* 264–266 (1985).

8 Needleman, P.; Greenwald, J.E.: Atriopeptin: a cardiac hormone intimately involved in fluid electrolyte, and blood pressure homeostasis. New Engl. J. Med. *314:* 828–834 (1986).

9 Rascher, W.; Tulassay, T.: Hormonal regulation of water metabolism in children with nephrotic syndrome. Kidney int. *32:* suppl. 21, pp. 83–89 (1987).

10 Rascher, W.; Tulassay, T.; Lang, R.E.: Atrial natriuretic peptide in plasma of volume-overloaded children with chronic renal failure. Lancet *ii:* 303–305 (1985).

11 Rascher, W.; Tulassay, T.; Seyberth, H.W.; Himbert, U.; Lang, U.; Schärer, K.: Diuretic and hormonal responses to head-out water immersion in nephrotic syndrome. J. Pediat. *109:* 609–614 (1986).

12 Richards, A.M.; Ikram, H.; Yandle, T.G.; Nicholls, M.G.; Webster, M.W.I.; Espiner, E.A.: Renal, haemodynamic, and hormonal effects of human alpha atrial natriuretic peptide in healthy volunteers. Lancet *i:* 545–549 (1985).

13 Sagnella, G.A.; Markandu, N.D.; Shore, A.C.; MacGregor, G.A.: Effects of changes in dietary sodium intake and saline infusion on immunoreactive atrial natriuretic peptide in human plasma. Lancet *ii:* 1208–1211 (1985).

14 Seifter, J.L.; Skorecki, K.L.; Stievelmen, J.C.; Haupert, G.; Brenner, B.M.: Control of extracellular fluid volume and pathophysiology of edema formation; in Brenner, Rector, The kidney, pp. 343–384 (Saunders, Philadelphia 1985).
15 Tulassay, T.; Rascher, W.; Lang, R.E.; Seyberth, H.W.; Schärer, K.: Atrial natriuretic peptide and other vasoactive hormones in nephrotic syndrome. Kidney int. *31:* 1391–1395 (1987).
16 Weidmann, P.; Hasler, L.; Gnädinger, M.P.; Lang, R.E.; Uehlinger, D.E.; Shaw, S.; Rascher, W.; Reubi, F.C.: Blood levels and renal effects of atrial natriuretic peptide in normal man. J. clin. Invest. *77:* 734–742 (1986).
17 Wilkins, M.R.; Wood, J.A.; Adu, D.; Lote, C.J.; Kendall, M.J.; Michael, J.: Change in plasma immunoreactive atrial natriuretic peptide during sequential ultrafiltration and haemodialysis. Clin. Sci. *71:* 157–160 (1986).
18 Yamaji, T.; Ighibashi, U.; Takaku, F.: Atrial natriuretic factor in human blood. J. clin. Invest. *76:* 1705–1709 (1985).

Wolfgang Rascher, MD, Universitätsklinik für Kinder- und Jugendmedizin der GHS, Hufelandstrasse 55, D–4300 Essen 1 (FRG)

Contr. Nephrol., vol. 67, pp. 9–16 (Karger, Basel 1988)

Recent Advances in Prostanoids and Renal Disease

T.J. Beattie[1]

Royal Hospital for Sick Children, Glasgow, UK

Despite the involvement of many disciplines in the field of renal prostanoids, understanding of the role of prostanoids, i.e. prostaglandins and thromboxanes, in the pathophysiology of a variety of renal disease is as yet incomplete.

The range of renal disease in which disturbances in prostanoids have been implicated is shown in table I. Full discussion of each of these areas is outwith the scope of this review and the following discussion will be limited to two areas only. Firstly, that of thrombotic microangiopathy, particularly the acute childhood haemolytic uraemic syndrome (HUS), and, secondly, some aspects of prostanoids in renal transplantation. Reference will be made only to published data on prostanoids since data on leukotrienes and other arachidonic acid metabolites remain very limited and are restricted in general to experimental models.

Haemolytic Uraemic Syndrome

The ability of normal plasma to support the production of prostacyclin-like activity from vascular tissue in vitro was first reported in 1978 [1]. Subsequent attempts at characterising this activity have revealed the presence of a factor with a molecular weight of $< 10{,}000$, which apparently acts both by enhancing the release of endogenous arachidonic acid from cell membranes and by preventing self-deactivation of the vascular cyclo-oxygenase enzyme system [2].

[1] I wish to thank Mrs. Monaghan for excellent secretarial assistance.

Table I. Prostanoids and renal disease

Diabetes insipidus	Glomerulonephritis and nephrotic syndrome
Bartter's syndrome	Renal and essential hypertension
Acute renal failure	Urolithiasis and hypercalcaemia
Hepato-renal syndrome	Sub-total nephrectomy
Chronic renal failure	Diabetic nephropathy
HUS/TTP	Renal transplantation

Remuzzi et al. [3] were the first group to demonstrate that this plasma prostacyclin-supporting activity (PSA) was decreased in both the HUS, and thrombotic thrombocytopenic purpura (TTP).

In view of the nature of the in vitro assay, diminished PSA could be explained by several reasons occurring singly or in combination. Firstly, by a deficiency of some plasma factor(s) as mentioned above; secondly, by the presence of plasma inhibitory activity either to the production or to the effect of prostacyclin, and, thirdly, by rapid degradation of prostacyclin, possibly due to a lack of stabilising factor [4]. Subsequent reports in acute childhood HUS [5–9] have, in general, been confirmatory. However, some groups have failed to demonstrate this abnormality in typical cases of HUS [10], and others have demonstrated the abnormality in PSA in atypical cases only [11]. With regard to the aetiology of the diminished PSA, inhibitory activity has been documented by several investigators [7, 8, 11]. It has been suggested that at least part of this inhibitory activity may be secondary to the presence of circulating lipid peroxides [11], which are known to inhibit prostacyclin synthesis. Further characterisation of this inhibitory activity is obviously required.

The finding of diminished PSA in the family members of these patients has been reported by several groups [8, 9, 12]. The relevance of these findings, however, is not clear but what is evident is that diminished PSA is not inconsistent with longevity. It would seem, therefore, that this prostanoid defect may be demonstrated fairly consistently in the HUS, but like the syndrome itself, the defect seems to be heterogenous. It should be appreciated, however, that PSA, rather than be dependent on one mechanism, may well involve several integrated steps, and the biological test systems used to date may be sensitive to only one of these. It is probably, therefore, no suprise to find that this defect in PSA is not specific to either HUS or TTP and has been demonstrated in a variety of different conditions as diverse as sickle cell anaemia [13], acute glomerulonephritis [6], and Henoch-Schönlein purpura [14].

The question is whether these data are of any relevance in formulating new or consolidating existing hypotheses on the pathophysiology of the thrombotic microangiopathies. Whatever the trigger or triggers are in HUS and TTP the initial mechanism is that of endothelial damage, leading ultimately to microvascular thrombosis. The bulk of published data suggests that the vascular prostanoid abnormality is of secondary importance either because it is induced by endothelial damage and/or lipid peroxidation or if the family studies are of significance the defect may merely amplify the platelet aggregatory response to some other stimuli.

The logical extension of this proposal is that therapeutic measures, particularly plasma exchange, may be successful for reasons other than correction of this defect, such as removal of endothelial toxic factors, or platelet aggregatory agents or by improvement in plasma anti-oxidant activity.

Renal Transplantation

Acute Rejection

In acute renal allograft rejection, increased platelet deposition and reduced graft perfusion are early features which may potentiate and in part be responsible for graft destruction by infarction. In experimental allograft rejection the most consistent finding is that of an increase in graft thromboxane production [15]; in addition, the haemodynamic changes associated with the rejection reaction may be partially responsive to thromboxane synthetase inhibition [16].

The clinical data are rather scanty. An increase in human renal allograft tissue slice production of 6-keto-PGF_1-alpha has been reported [17], but these grafts were at a very advanced stage in the rejecting process. Increases in both urinary PGE_2 concentration [18] and urinary thromboxane B excretion [19–21] have been proposed as sensitive and early indicators of acute rejection. The increase in urinary thromboxane B_2 excretion was not affected by other events, such as acute infection, acute tubular necrosis or by ciclosporin therapy, but increased urinary excretion may be seen in systemic deep venous thrombosis or pulmonary thrombo-embolism.

Urinary thromboxane excretion may, therefore, be a useful clinical tool in the peri-transplant period, particularly in view of the fact that symptoms associated with acute rejection are frequently absent in ciclosporin-treated patients. If urinary PGE_2 is being used as a monitor, some caution must be exercised since increased excretion of this prostanoid has been demonstrated in grafts recovering from acute tubular necrosis [22].

Ciclosporin

Since the introduction of ciclosporin, a large volume of data have accumulated on its mode of action and nephrotoxicity. In relation to the nephrotoxic effects, there have been a number of suggestions that some or all of these effects may be mediated through alteration in renal prostanoid metabolism. The adverse effects in which abnormalities in prostanoids have been implicated are those of functional relevance, i.e. reduction in glomerular filtration rate, hyperkalaemic metabolic acidosis and systemic hypertension, and morphological features such as arteriolopathy [23], which is similar to that seen in thrombotic microangiopathy and that of the capillary microangiopathy with or without features of HUS.

Unfortunately, despite the extensive experimental data, there seems to be no clear understanding of the action of ciclosporin on prostanoid metabolism but it may be of value to give some details on a sample of these reports. Neild et al. [24] demonstrated that plasma from rabbits treated with ciclosporin had a limited ability to support prostacyclin. This was felt to be due to a deficiency of the prostacyclin-stimulating factor. In addition, this group have recently reported a direct inhibitory effect of ciclosporin on prostacyclin production by human vascular endothelium [25]. In the rat model there have been reports on both decreased [26] and increased [27] prostanoid synthesis. A consistent finding in the animal models is that of an increase in the plasma renin activity in association with a decrease in renal blood flow and glomerular filtration rate, both with acute and chronic administration of ciclosporin, and these effects may [28, 29] or may not [30] be enhanced by cyclo-oxygenase inhibition. It is of interest that in the study by Murray et al. [29] the renal vascular resistance could be significantly reduced by alpha blockade, and more recent data [30, 31] suggest that calcium channel blockers may be even more effective in reversing this increase in renal vascular resistance. Two points about the experimental data are worth clarifying: firstly, in the great majority of studies the experimental animals were surgically intact, and, secondly, the doses of ciclosporin used varied between 2 and 20 times the immunosuppressive dose required. It is well recognised that doses of this order may induce anorexia and diarrhoea which may lead to a fall in extracellular fluid volume and a secondary rise in plasma renin activity. In view of this and other evidence suggesting that the effects of ciclosporin in experimental animals are dose, time and species dependent, it is not surprising that the results, apart from the finding of increased renal vascular resistance, are difficult to assess.

The clinical data are much less extensive. In contrast to the experimental models, the plasma renin activity is depressed [32], in association with an increase in the extracellular fluid volume, which is responsive to salt depletion [33]. Decreased urinary PGE_2 excretion has been documented in renal allograft recipients [34, 35] and studies in non-allografted patients have also revealed ciclosporin-induced changes in urinary prostanoid excretion [36, 37].

Comparison of data, pre- and post-conversion from ciclosporin therapy are of interest and Curtis et al. [38] and Jorkasky et al. [39] have shown a decrease in renal vascular resistance and increased renal blood flow with no changes in urinary prostanoid excretion following conversion to azathioprine. In addition to these data, there have been no reports on additive nephrotoxicity with the use of cyclo-oxygenase inhibitors in clinical studies.

As far as the morphological findings are concerned it would seem that there are no data either way to implicate a disturbance in prostanoid metabolism in the development of the arteriolopathy, but as this seems to be a feature of chronic ciclosporin therapy, a direct cytotoxic effect is probably more relevant. With regard to the microangiopathy, there have been very few reports of this occurring in ciclosporin-treated patients in the absence of acute vascular rejection [40, 41] and there are only 3 patients reported to date who developed de novo HUS after *renal* transplantation [42]. In addition, those patients in whom the primary diagnosis is HUS do not seem to be at an increased risk of recurrence after renal transplantation, despite immunosuppression with ciclosporin [43].

With regard to the functional adverse effects, there seems little doubt that the diminished renal blood flow and glomerular filtration rate are due to renal arteriolar vasoconstriction. A secondary increase in proximal tubular sodium reabsorption has been demonstrated in both experimental [44] and clinical studies [45, 46]. This would have the effect of reducing distal tubular sodium delivery and of increasing extracellular fluid volume with a secondary reduction in plasma renin. Both of these effects could lead to diminished distal tubular potassium and hydrogen ion secretion and hypertension. The clinical data do not support a role for either a reduction in vasodilatory prostanoids or an increase in vasoconstrictor prostanoids as the primary cause of the arteriolar vasoconstriction. These data, however, do not exclude the presence of a blunted prostanoid response to alternative vasoconstrictor stimuli such as enhanced responsiveness to circulating catecholamines and/or sympathetic innervation or, more likely, a direct effect of ciclosporin on vascular smooth muscle.

References

1 Mac Intyre, D.E.; Pearson, J.D.; Gordon, J.L.: Localisation and stimulation of prostacyclin production in vascular cells. Nature, Lond. *271:* 549–551 (1978).

2 Deckmyn, H.; Zoja, C.; Arnout, J.; Todisco, A.; Vanden Bulcke, F.; D'Hondt, L.; Hendrickx, N.; Gresele, P.; Vermylen, J.: Partial isolation and function of the prostacyclin regulating plasma factor. Clin. Sci. *69:* 383–393 (1985).

3 Remuzzi, G.; Misiani, R.; Marchesi, D.; Livio, M.; Mecca, G.; de Gaetano, G.; Donati, M.B.: Haemolytic-uraemic syndrome: deficiency of plasma factor(s) regulating prostacyclin activity? Lancet *ii:* 871–872 (1978).

4 Wu, K.K.; Hall, E.R.; Rossi, E.C.; Papp, Alc.: Serum prostacyclin binding defects in thrombotic thrombocytopenic purpura. J. clin. Invest. *75:* 168–174 (1985).

5 Defreyn, G.; Proesmans, W.; Machin, S.J.; Lemmens, F.; Vermylen, J.: Abnormal prostacyclin metabolism in the haemolytic uraemic syndrome: equivocal effect of prostacyclin infusions. Clin. Nephrol. *18:* 43–49 (1982).

6 Chamone, D.A.; Proesmans, W.C.; Monnens, L.A.; Binda ki Muaka, P.; Vermylen, J.G.: Reversible deficient prostacyclin release in childhood hemolytic uremic syndrome. Int. J. pediat. Nephrol. *3:* 13–16 (1982).

7 Stuart, M.J.; Spitzer, R.E.; Coppe, D.: Abnormal platelet and vascular prostaglandin synthesis in an infant with hemolytic uremic syndrome. Pediatrics, Springfield *71:* 120–123 (1983).

8 Turi, S.; Beattie, T.J.; Belch, J.J.F.; Murphy, A.V.: Disturbances of prostacyclin metabolism in children with hemolytic-uremic syndrome and in first degree relatives. Clin. Nephrol. *25:* 193–198 (1986).

9 Seigler, R.L.; Smith, J.B.; Lynch, M.B.; Mohammad, S.F.: In vitro prostacyclin production in the hemolytic-uremic syndrome. West. J. Med. *144:* 165–168 (1986).

10 Rizzoni, G.; Pavanello, L.; Claris-Appiani, A.; Edefonti, A.; Facchin, P.; Franchini, F.; Gussmano, R.; Imbasciati, E.; Perfumo, F.; Remuzzi, G.: Treatment of children with hemolytic uremic syndrome (HUS) with plasma. A multicenter controlled trial (Abstract). Helv. paediat. Acta *41:* 114 (1986).

11 Levin, M.; Elkon, K.B.; Nokes, T.J.C.; Buckle, A.M.; Dillon, M.J.; Hardisty, R.M.; Barratt, T.M.: Inhibitor of prostacyclin production in sporadic haemolytic uraemic syndrome. Archs Dis. Childh. *58:* 703–708 (1983).

12 Jorgensen, K.A.; Pedersen, R.S.: Familial deficiency of prostacyclin production stimulating factor in the hemolytic uremic syndrome of childhood. Thromb. Res. *21:* 311–315 (1981).

13 Stuart, M.J.; Sills, R.H.: Deficiency of plasma prostacyclin or PGI2 regenerating ability in sickle cell anaemia. Br. J. Haemat. *48:* 545–550 (1981).

14 Turi, S.; Belch, J.J.F.; Beattie, T.J.; Forbes, C.D.: Abnormalities of vascular prostaglandins in Henoch-Schönlein purpura. Archs Dis. Childh. *61:* 173–177 (1986).

15 Gibbons, C.P.; Wiley, K.N.; Lindsey, N.J.; Fox, M.; Beck, S.; Slater, D.N.; Preston, F.E.; Brown, C.B.; Raftery, A.T.: Cortical and vascular prostaglandin synthesis during renal allograft rejection in the rat. Transplantation *43:* 472–478 (1987).
16 Coffman, T.M.; Yarger, W.E.; Klotman, P.E.: Functional role of thromboxane production by acutely rejecting renal allografts in rats. J. clin. Invest. *75:* 1242–1248 (1985).
17 Leithner, C.; Sinzinger, H.; Silberbaver, K.; Wolf, A.; Stummvoll, H.K.; Pinggera, W.: Enhanced prostacyclin synthesis in acute human kidney transplant rejection. Proc. Eur. Dial. Transplant. Ass. *17:* 424–428 (1980).
18 Thompson, D.; Rowe, D.J.F.; Briggs, J.D.: Measurement of urine prostaglandin E2 as a predictor of acute renal transplant rejection: preliminary findings. Ann. clin. Biochem. *22:* 161–165 (1985).
19 Foegh, M.L.; Alijani, M.R.; Helfrich, B.S.; Khirabadi, B.S.; Lim, K.; Ramwell, P.W.: Lipid mediators in organ transplantation. Transplant. Proc. *18:* 20–24 (1986).
20 Steinhaver, H.B.; Wilms, H.; Ruther, M.; Schollmeyer, P.: Clinical experience with urine TxB2 in acute renal allograft rejection. Transplant. Proc. *18:* 98–103 (1986).
21 Proppe, D.; Jentzen, V.; Uthgenannt, D.; Niedermayer, W.: Eicosanoid metabolites in the diagnosis of renal graft rejection (Abstract). Proc. Xth Int. Congr. Nephrol., 1987, p. 634.
22 Zawada, E.T.: Urinary prostaglandin E2 excretion after the resumption of renal function post renal transplantation. Int. J. artif. Organs *8:* 119–120 (1985).
23 Mihatsch, M.J.; Thiel, G.; Spichtin, H.P.; Oberholzer, M.; Brunner, F.P.; Harder, F.; Olivieri, V.; Bremer, R.; Ryffel, B.; Stocklin, E.; Torhorst, J.; Gudat, F.; Zollinger, H.U.; Loertscher, R.: Morphological findings in kidney transplants after treatment with cyclosporin. Transplant. Proc. *15:* 2821–2835 (1983).
24 Neild, G.H.; Rocchi, G.; Imberti, L.; Fumagalli, F.; Brown, Z.; Remuzzi, G.; Williams, D.G.: Effect of cyclosporin A on prostacyclin synthesis by vascular tissue. Thromb. Res. *32:* 373–379 (1983).
25 Brown, Z.; Neild, G.H.: Cyclosporin inhibits prostacyclin production by cultured human endothelial cells. Transplant. Proc. *19:* 1031–1033 (1987).
26 Stahl, R.A.K.; Kudelka, S.: Chronic cyclosporine treatment reduces prostaglandin E2 formation in isolated glomeruli and papilla of rat kidneys. Clin. Nephrol. *25:* S78–S82 (1986).
27 Coffman, T.M.; Carr, D.R.; Yarger W.E.; Klotman, P.E.: Evidence that renal prostaglandin and thromboxane production is stimulated in chronic cyclosporine nephrotoxicity. Transplantation *43:* 282–285 (1987).
28 Siegl, H.; Ryffez, B.; Petric, R.; Shoemaker, P.; Muller, A.; Donatsch, P.; Mihatsch, M.: Cyclosporine, the renin-angiotensin-aldosterone system, and renal adverse reactions. Transplant. Proc. *15:* 2719–2725 (1983).
29 Murray, B.M.; Paller, M.S.; Ferris, T.F.: Effect of cyclosporine administration on renal haemodynamics in conscious rats. Kidney int. *28:* 767–774 (1985).
30 Barros, E.J.G.; Boim, M.A.; Ajzen, H.; Ramos, O.L.; Schor, N.: Glomerular hemodynamics and hormonal participation on cyclosporine nephrotoxicity. Kidney int. *32:* 19–25 (1987).
31 Dieperink, H.; Leyssac, P.; Starklint, H.; Jorgenson, K.A.; Kemp, E.: Antagonist capacities of mifedepine, captopril, phenoxybenzamine, prostacyclin and indomethacin on cyclosporin A induced impairment of rat renal function. Eur. J. clin. Invest. *16:* 540–548 (1986).

32 Bantle, J.P.; Nath, K.A.; Sutherland, D.E.R.; Najarian, J.S.; Ferris, T.F.: Effects of cyclosporine on the renin-angiotensin-aldosterone system and potassium excretion in renal transplant recipients. Archs intern. Med. *145:* 505–508 (1985).

33 Bantle, J.P.; Boudreau, R.J.; Ferris, T.F.: Suppression of plasma renin activity by cyclosporine (Abstract). Clin. Res. *34:* 970A (1986).

34 Adu, D.; Lote, C.J.; Michael, J.; Turney, J.H.; Mcmaster, P.: Does cyclosporine inhibit renal prostaglandin synthesis? Proc. edta-era *21:* 969–972 (1984).

35 Stahl, R.A.K.; Kanz, L.; Kudelka, S.: Cyclosporine and renal prostaglandin E2 production. Ann. intern Med. *103:* 474 (1985).

36 Deray, G.; Le Hoang, P.; Caloub, P.; Baumelou, A.; Hornych, A.; Jacobs, C.: Effects of cyclosporin on plasma renin activity, aldosterone, prostaglandins, and catecholamines in patients with uveitis (Abstract). Proc. Xth Int. Congr. Nephrol., 1987, p. 457.

37 Kuhn, K.; Forstermann, U.; Frolich, J.C.; Vesterquist, O.; Green, K.; Brunkhorst, R.; Koch, K.M.; Hass, J.: Effects of cyclosporine on blood pressure and prostacyclin and thromboxane A2 production (Abstract). Proc. Xth Int. Congr. Nephrol., 1987, p. 231.

38 Curtis, J.J.; Luke, J.J.; Dubovsky, E.; Diethelm, A.G.; Whelchel, J.D.; Jones, P.: Cyclosporin in therapeutic doses increases renal allograft vascular resistance. Lancet *ii:* 477–479 (1986).

39 Jorkasky, D.; Audet, P.; Williams, S.; Grossman, R.; Conrad, M.: Cyclosporine induced nephrotoxicity: role of prostaglandins. Transplant. Proc. *19:* 1742–1744 (1987).

40 Neild, G.H.; Reuben, R.; Hartley, R.B.; Cameron, J.S.: Glomerular thrombi in renal allograft associated with cyclosporin treatment. J. clin. Path. *38:* 253–258 (1985).

41 Bonsib, S.M.; Smith, J.L.: Cyclosporin associated thrombotic microangiopathy: ultrastructural evidence of widespread endothelial damage (Abstract). Proc. Xth Int. Congr. Nephrol., 1987, p. 596.

42 Buren, D. van; Buren, C.T. van; Flechner, S.M.; Maddox, A.M.; Verani, R.; Kahan, B.D.: De novo hemolytic uremic syndrome in renal transplant recipients immunosuppressed with cyclosporine. Surgery *98:* 54–62 (1985).

43 Pirson, Y.; Leclercq, B.; Squifflet, J.P.; Alexandre, G.P.J.; Van Ypersele, C.: No recurrence of the hemolytic uremic syndrome on cyclosporin therapy after renal transplantation (Abstract). Proc. Xth Int. Congr. Nephrol., 1987, p. 633.

44 Dieperink, H.; Leyssac, P.; Starklint, H.; Kemp, E.: Nephrotoxicity of cyclosporin A.: A lithium clearance and micropuncture study in rats. Eur. J. clin. Invest. *16:* 69–77 (1986).

45 Propper, D.J.; Mackay, J.; Whiting, P.; Power, D.; Edward, N.; Catto, G.R.D.: Lithium clearance: a method of monitoring cyclosporin A nephrotoxicity (Abstract). Proc. Xth Int. Congr. Nephrol., 1987, p. 634.

46 Vincent, H.H.; Weimar, W.; Schalekamp, M.A.D.H.: Cyclosporine lowers fractional excretion of lithium and potassium in kidney transplant (Abstract). Proc. Xth Int. Congr. Nephrol., 1987, p. 491.

T.J. Beattie, MD, Royal Hospital for Sick Children, Yorkhill, Glasgow G3 8SJ (UK)

Contr. Nephrol., vol. 67, pp. 17–21 (Karger, Basel 1988)

Clinical Assessment of Renal and Systemic Prostanoid Activity[1]

H.W. Seyberth[2], *H. Schweer*

Children's Hospital, University of Heidelberg, FRG

A body of evidence continues to grow that renal prostanoids play an important role in kidney function under physiological and pathophysiological conditions. This includes glomerular as well as tubular function. However, in contrast to in vitro studies and despite many efforts, the analysis of prostanoids in vivo and particularly under clinical conditions still remains a delicate undertaking. Basically, this problem arises from the following factors: First, active prostanoids have a very short biological half-life and high potency. The resulting low levels in biological fluids require extreme analytical sensitivity. Second, a broad spectrum of prostanoid derivatives with minor structural variation, yet with major differences in biological activity, have already been identified and still more are waiting to be discovered. Third, the biosynthetic capacity of tissue greatly exceeds the actual rate of synthesis in vivo. A variety of stimuli, e.g. plain mechanical manipulation, will trigger the arachidonate-prostanoid cascade leading to ex vivo contamination of a sample which contains cells with the capability of synthesizing prostanoids. These capacity-related measurements have often been misinterpreted as reflecting endogenous biosynthesis. Likewise, determinations of peripheral plasma concentrations of primary prostanoids or their spontaneously formed hydration products have turned out to be an unreliable approach to assess prostanoid activity under clinical condition (fig. 1) [3]. Moreover, because of the rapid metabolism in lung,

[1] The author's studies referred to in this paper were supported by grants from the Deutsche Forschungsgemeinschaft (Se 263).

[2] H.W.S. is a Hermann and Lilly Schilling Professor for Medical Research.

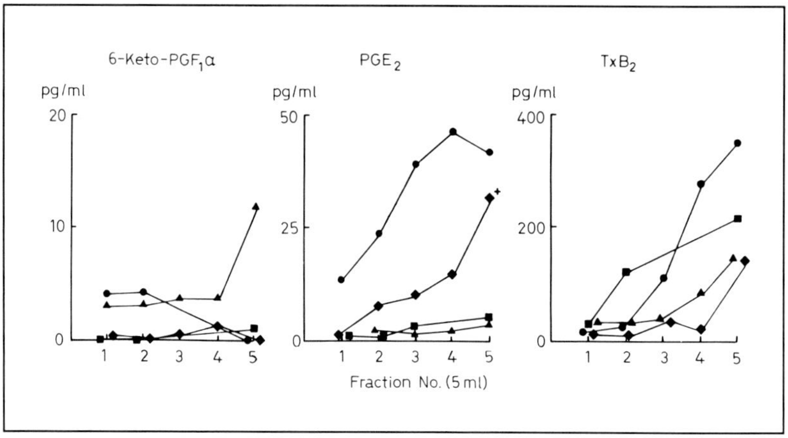

Fig. 1. Peripheral plasma levels of 6-keto-PGF$_1\alpha$, PGE$_2$ and TxB$_2$ in 4 healthy volunteers during a fractional blood collection. + = Needle was clogged [reproduced with modification from ref. 3].

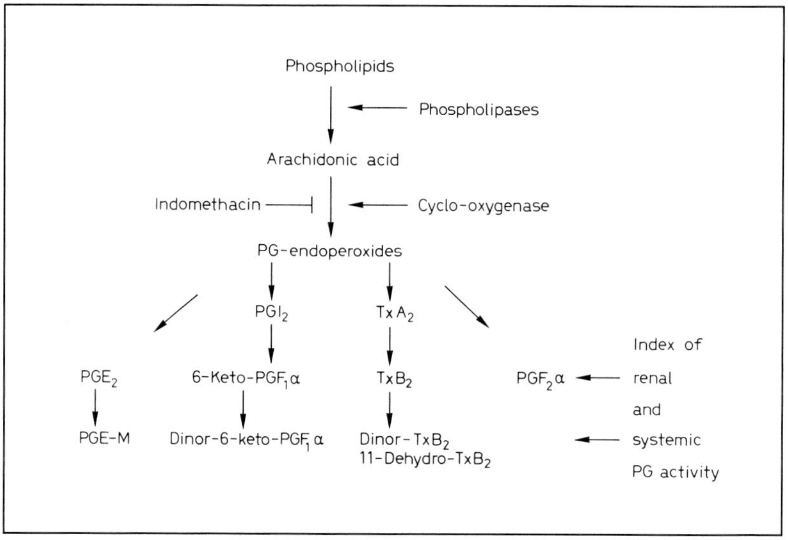

Fig. 2. Cyclo-oxygenase pathway of the arachidonate cascade leading to the primary prostanoids and their enzymatic metabolites in urine. The excretion rate of the former are considered as the index of renal and that of the latter as the index of systemic prostanoid activity.

kidney and liver and the universal synthesis capability in the body, peripheral plasma concentrations of prostanoids have by no means any relationship with renal prostanoid activity.

After the discovery of primary prostanoids in human urine [2], their urinary excretion rates have frequently been used for in vivo assessment of renal prostanoid activity. This approach appears to be justified because urine normally represents an almost cell-free medium, avoiding the ex vivo artifact, and systemically administered or generated prostanoids are primarily excreted as metabolites [1].

Since then, the simultaneous determination of urinary excretion rates of primary prostanoids including their hydration products and the enzymatically formed prostanoid metabolites has frequently been employed to assess renal and systemic prostanoid activity, respectively (fig. 2). Fortunately, there appeared to be no significant influence on these excretion rates by chronic changes in kidney function including glomerular filtration rate, urinary osmolality, pH and urine flow [unpubl. observations].

Employing this methodological approach for the assessment of renal and systemic prostanoid activity and using highly sophisticated mass spectrometric methodology, we were able to describe a congenital disease entity, in which renal and systemic PGE_2 synthesis is selectively stimulated (fig. 3). After the treatment with indomethacin, a potent inhibitor of prostanoid synthesis, was proven to be the most effective therapy in these children [4], it became apparent that this PGE_2 hyperactivity is causally related to the pathomechanism of this disease and led to the introduction of the term hyperprostaglandin E_2 syndrome [5]. In table I the symptoms and signs of the hyperprostaglandin E_2 syndrome are related to the increased renal and systemic PGE_2 overproduction, as determined by the urinary excretion rates of PGE_2 and PGE-M, respectively.

In the future, it will become more apparent that other prostanoids might function as mediators of pathophysiological processes within the kidney as well. In this respect TxA_2 has already been implicated in a variety of renal diseases, such as vascular host-graft reaction after kidney transplantation, lupus nephritis, hydronephrosis, and ciclosporin nephrotoxicity. However, more caution should be taken by performing determinations of prostanoids in urine. Many presently used immunoassays have to be validated by chromatographic methods, by use of multiple antisera to identify cross-reactivities, and by comparison with a reference method, preferably with gas chromatography-mass spectrometry [1]. Moreover, with the increasing realization of the complexity of this field and knowl-

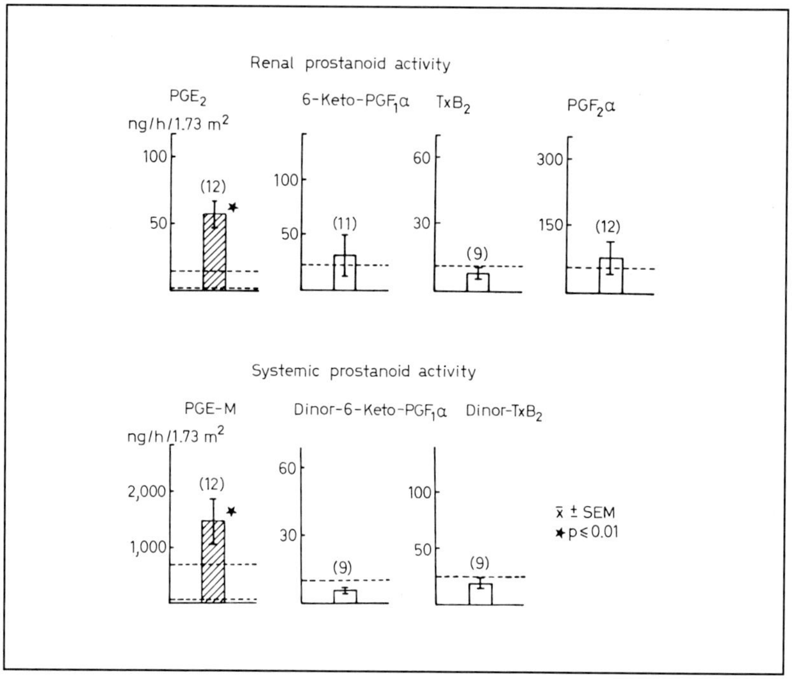

Fig. 3. Urinary excretion rates of primary prostanoids (renal prostanoid activity) and enzymatic metabolites (systemic prostanoid activity) in infants with hyperprostaglandin E_2 syndrome. For technical reasons not all 7 prostanoid excretion rates were obtained in all 12 patients.

Table I. Symptoms and signs of infants with hyperprostaglandin E_2 syndrome in relation to increased renal and systemic PGE_2 synthesis

Renal effects (urinary PGE_2 ↑)	polyhydramnios (prematurity) polyuria with isosthenuria hypokalemic alkalosis hypercalciuria with nephrocalcinosis
Systemic effects (urinary PGE-M ↑)	fever diarrhea osteopenia normotension despite high renin and ADH activity prone to infection failure to thrive

edge of how the different metabolic pathways may interfere with one another, the determination of a single prostanoid becomes less meaningful. For this reason a great deal of effort is at present going into the development of simultaneous monitoring of a spectrum of prostanoid derivatives. By these means, one might be more able to identify the actual mediator of a pathophysiological process and its place of synthesis, e.g. of renal or nonrenal origin.

References

1 Catella, F.; Nowak, J.; FitzGerald, G.A.: Measurement of renal and non-renal eicosanoid synthesis. Am. J. Med. *81:* suppl. 2B, pp. 23–29 (1986).

2 Frölich, J.C.; Wilson, T.W.; Sweatman, B.J.; Smigel, M.; Nies, A.S.; Carr, K.; Watson, J.T.; Oates, J.A.: Urinary prostaglandins. Identification and origin. J. clin. Invest. *55:* 763–770 (1975).

3 Schweer, H.; Kammer, J.; Kühl, P.G.; Seyberth, H.W.: Determination of peripheral plasma prostanoid concentration. An unreliable index of 'in vivo' prostanoid activity. Eur. J. clin. Pharmacol. *31:* 303–305 (1986).

4 Seyberth, H.W.; Rascher, W.; Schweer, H.; Kühl, P.G.; Mehls, O.; Schärer, K.: Congenital hypokalemia with hypercalciuria in preterm infants. A hyperprostaglandinuric tubular syndrome different from Bartter syndrome. J. Pediat. *107:* 694–701 (1985).

5 Seyberth, H.W.; Königer, S.J.; Rascher, W.; Kühl, P.G.; Schweer, H.: Role of prostaglandins in hyperprostaglandin E syndrome and in selected tubular disorders. Pediat. Nephrol. *1:* 491–497 (1987).

Prof. H.W. Seyberth, MD, Universitäts-Kinderklinik, Im Neuenheimer Feld 150, D-6900 Heidelberg (FRG)

Contr. Nephrol., vol. 67, pp. 22–27 (Karger, Basel 1988)

Rheological Mechanisms in Postischemic Acute Renal Failure

G. Losonczy

Semmelweis University Medical School, Department of Pathophysiology, Budapest, Hungary

Vascular Obstruction

Recent observations indicate the significance of partial or complete vascular obstruction over functional vasoconstriction as the mechanism of early decrease of renal blood flow in experimental postischemic acute renal failure [1–3]. Light-microscopic examination reveals cellular aggregates built up mostly from red blood cells both in the cortex and medulla [4]. The inner strip of the outer medulla is the site of most severe vascular congestion. Single medullary capillaries outside the bundles of the vasa recta fill up tightly with aggregated erythrocytes and their diameter becomes increased manyfold during the first minutes of postischemic reperfusion. In these capillaries blood flow will cease again shortly after the release of renal artery occlusion.

The partial or complete vascular obstruction as well as the marked increase of capillary volume result in multiple functional consequences. Postischemic glomeruli of all cortical layers produce less filtrate as compared to normal conditions, but the most severe depression of filtration is observed in deep and juxtamedullary glomeruli drained by medullary capillaries affected most severely by congestion after ischemia [5].

Furthermore, the congested capillaries hinder tubular flow and, indirectly, glomerular filtration by exerting external compression to the tubules [6].

Capillary volume in the inner strip is increased by 250–300%. The same in the cortex is about 25% [2]. Cellular aggregates also form in the glomerular capillaries [7]. This mechanism may be hidden behind the fact that at least 1/3 of the postischemic glomeruli stop filtration because of primary glomerular injury [8].

The question was put whether postischemic kidney function improves by preventing the formation of, or flushing out the cellular aggregates causing vascular obstruction [1]? Two rheological procedures were tried through which – at least directly – tubular ischemic damage could certainly not be lessened: the formation of aggregates was prevented by anemic reperfusion (hematocrit value 31% compared with 44% in untreated ischemic rats), and the aggregates already formed were flushed out by high-pressure reperfusion (mean blood pressure was 30 mm Hg higher than in untreated ischemics). Actually, these maneuvers hardly attenuated tubular injury; however, renal blood flow increased from 50 to 100% of control and GFR from 20 to 60% of the normal value. Fifty percent of tubular reabsorptive capacity lost had been recovered, too. The efficacy of these experimental treatments allow two conclusions to be drawn: (1) early postischemic decrease of renal blood flow is due mostly to obstruction of vessels by cellular aggregates; (2) the vascular obstruction is responsible for about 50% of early postischemic glomerular and tubular dysfunction [1].

Possible Mechanisms of Renal Postischemic Cellular Aggregation

'Simple erythrocyte aggregation' [1] and intrarenal coagulation [9–16] arise as mechanisms responsible for cellular obstruction of vessels in the postischemic kidney. During complete cessation of blood flow, vascular permeability is increased, plasma leaves the vascular space and local hemoconcentration results. The more severe the regional hemoconcentration the more incomplete the reperfusion after the release of the arterial clamps. This way, incomplete reflow maintains relative tissue ischemia leading again to enhanced capillary permeability. Ischemia, hemoconcentration and high resistance to flow due to increased viscosity form the vicious cycle leading to the regional no-reflow phenomenon. Simple erythrocyte aggregation works without the formation of any adhesive (e.g. fibrin) between cells. Mason et al. [1] claimed that fibrin and platelet aggregates had no role in postischemic vascular obstruction, since heparin and aspirin treatments were ineffective against it. Intrarenal coagulation was not examined and the actual effectiveness of these anticoagulant measures against postischemic clotting were not controlled.

Light- and electron-microscopic studies do not always identify the presence of clots in the postischemic kidney [4, 7]. A number of publications, however, support intrarenal coagulation as a characteristic feature in both human [9–11] and experimental renal ischemia [12–16]. The controversy may be explained partly by the induction of ischemia of diverse severity, but mostly by the especially high fibrinolytic activity of the kidneys [17]. In postischemic [13] and other forms of intrarenal coagulation [17] fibrinolysis is also activated and may dissolve fibrin before fixation.

These considerations formed the rationale for the re-examination of intrarenal coagulation in the rat model of human reversible shock kidney [14, 15]. The right renal artery was completely occluded for 45 min after left uninephrectomy performed 3 weeks earlier. Intrarenal thrombus formation was studied after 30 min reperfusion using ^{51}Cr-labelled autologous platelets and ^{125}I-labelled human fibrinogen. Control and postisch-

Table I. Intrarenal coagulation and glomerular filtration rate (GFR) after 45 min complete right renal artery occlusion[1] and 30 min reperfusion in the rat ($\bar{x} \pm$ SE)

	Cortex			Medulla		
	control (n = 9)	postischemia		control (n = 9)	postischemia	
		untreated (n = 8)	heparin (n = 8)		untreated (n = 8)	heparin (n = 8)
Platelets × 10^7/g tissue	8.2 ± 1.1	16.3 ± 1.7[a]	17.1 ± 4.0[b, n.s.]	10.0 ± 1.1	53.9 ± 7.8[a]	43.2 ± 11.0[b, n.s.]
Fibrin µg/g tissue	0.0[2]	130.3 ± 15.0	75.0 ± 11.0[b]	0.0	94.0 ± 13.0	15.0 ± 6.1[c]

GFR in control animals was 0.61 ± 0.04 ml/min/100 g body weight. GFR in postischemic untreated animals was 0.02 ± 0.01 ml/min/100 g, $p < 0.001$ as compared to controls. GFR in heparin-treated animals was 0.03 ± 0.01 ml/min/100 g BW, $p < 0.001$ as compared to controls and not significantly different from untreated ischemics.

[a] $p < 0.001$; [b] $p < 0.02$ as compared to controls; [c] $p < 0.001$; n.s. = not significant as compared to untreated ischemics.

[1] Three weeks earlier left uninephrectomy was performed.

[2] The calculation of the amount of ischemia-induced fibrin included the subtraction of sedimentable radioactivity inevitably present due to tissue processing [18]. Therefore, control tissues are considered to be free of fibrin.

emic cortical and medullary tissue samples were homogenized and centrifuged. Labelled fibrin was separated from labelled fibrinogen and other fibrinogen-fibrin derivatives (fibrin monomers, fibrin degradation products, etc.) on the basis of its particulate, water insoluble and sedimentable nature [18]. Blood ^{51}Cr activity and platelet count and plasma-clottable ^{125}I activity and fibrinogen concentration made possible the quantitative detection of platelets and fibrin deposited in grams of cortical and medullary tissue [15]. The results are summarized in table I.

The postischemic cortex contained approximately twice as much platelets as controls and 130 μg/g fibrin. The postischemic medulla contained about 5 times as much platelets as control tissue and on average 94 μg/g fibrin.

After the 30-min reperfusion period, the cross-section of the kidneys unequivocally shows medullary hyperemia, which means that intrarenal coagulation and erythrocyte aggregation appear simultaneously. Fibrin or aggregated platelets may be the adhesive between red blood cells and endothelial cells as well. Therefore, the effectiveness of high-dose heparin administration (500 U/100 g body weight, i.v. 5 min prior to renal artery occlusion, Chemical Works of Richter-Gedeon, Ltd., Budapest) was tested in the prevention of (1) postischemic intrarenal coagulation, and (2) postischemic decrease of glomerular filtration rate.

Heparin treatment had no influence at all on postischemic platelet accumulation (table I). Medullary fibrin formation was, in the large part, prevented; however, the amount of cortical fibrin deposition proved to be attenuated only by high-dose heparin. The glomerular filtration rate was less than 5% of control in both untreated and treated ischemic animals. These observations relate to three possible pathogenetic events: (1) erythrocyte aggregation moves on without fibrin as adhesive in the medulla; (2) however, medullary platelet aggregation may contribute to vascular obstruction, and (3) the postischemic fibrin and platelet deposits in the cortex may impair glomerular function and diminish cortical blood flow.

The cortical clot as a mass appears to be insignificant, but makes one think about the respective part of the cortical vasculature as the location of heparin-resistant fibrin formation (and platelet deposition)? Glomerular capillaries are known to be disposed to coagulation in several diseases which occur with thrombophilia: preeclampsia [19, 20], generalized Shwartzman reaction [21], postpartum acute renal failure [22], hemolytic-uremic syndrome, and thrombotic-thrombocytopenic purpura [23]. Moreover, heparin was actually found to be ineffective in the prevention of glomerular coagulation in the hemolytic-uremic syndrome [23] and in experimental renal ischemia induced in rabbits by 3–4 h of complete renal artery occlusion [16]. If the minute amount of heparin-resistant fibrin per gram cortical tissue after 45 min ischemia was localized to the glomeruli, postischemic kidney function was impaired considerably.

The nature of the triggering stimulus for postischemic clotting is largely unknown. One possibility is that reactive oxygen species released during the early reoxygenation period [24] make glomerular mesangial cells produce prostaglandins [25]. Glomeruli produce PGI_2 and TxA_2 [26]. Reactive oxygen species are known to inactivate PGI_2 synthetase, while TxA_2 synthetase remains intact [27]. Therefore, during the postischemic period studied, the critical balance between PGI_2 and TxA_2 may be lost and the proaggregator and vasoconstrictor TxA_2 may dominate. The importance of this substance is supported by the beneficial effect of imidazole, a selective inhibitor of TxA_2 synthetase, in the early postischemic phase [28].

References

1 Mason, J.; Welsch, J.; Torhorst, J.: The contribution of vascular obstruction to the functional defect that follows renal ischemia. Kidney int. *31:* 65–71 (1987).

2 Wolgast, M.; Karlberg, L.; Kallskog, Ö.; Norlén, B.J.; Nygren, K.; Öjteg, G.: Hemodynamic alterations in ischemic acute renal failure. Nephron *31:* 301–303 (1982).

3 Hansson, R.; Jonsson, O.; Lundstam, S.; Pettersson, S.; Schersten, T.; Waldenstrom, J.: Effects of free radical scavengers on renal circulation after ischemia in the rabbit. Clin. Sci. *65:* 605–610 (1983).

4 Torhorst, J.; Rougemont, D. de; Brunner, F.P.; Thiel, G.: Morphology of the renal medulla in ischemic acute renal failure in the rat. Nephron *31:* 296–300 (1982).

5 Mason, J.; Torhorst, J.; Welsch, J.: Role of the medullary perfusion defect in the pathogenesis of ischemic renal failure. Kidney int. *26:* 283–293 (1984).

6 Thiel, G.; Rougemont, D. de; Torhorst, J.; Kaufmann, A.; Peters-Haefeli, L.; Brunner, F.P.: Importance of tubular obstruction and its prevention in ischemic acute renal failure in the rat; in Leaf, Renal pathophysiology, pp. 223–238 (Raven Press, New York 1980).

7 Frega, N.S.; DiBona, D.R.; Gürtler, B.; Leaf, A.: Ischemic renal injury. Kidney int. *10:* S17–S25 (1976).

8 Conger, J.D.; Robinette, J.B.; Kelleher, S.P.: Nephron heterogeneity in ischemic acute renal failure. Kidney int. *26:* 422–429 (1984).

9 Clarkson, A.R.; McDonald, M.K.; Foster, V.; Cash, J.D.; Robson, J.S.: Glomerular coagulation in acute ischemic renal failure. Q. Jl Med. *39:* 585–599 (1970).

10 Pirani, C.L.: Coagulation and renal disease; in Remuzzi, Glomerular injury, 300 years after Morgagni, pp. 119–138 (Wichtig Editore, Milano 1983).

11 Kincaid-Smith, P.: Participation of intravascular coagulation in the pathogenesis of glomerular and vascular lesions. Kidney int. *7:* 242–253 (1975).
12 Wardle, E.N.: Intravascular coagulation in experimental acute renal failure: assessment by radio-fibrinogen technique. Thromb. Diath. Haemorrh. *29:* 579–591 (1973).
13 Losonczy, G.; Hársing, L.: Simultaneous activation of coagulation and fibrinolysis after bilateral renal ischemia in rats. Nephron *32:* 180–184 (1982).
14 Losonczy, G.: Ischemia induced intrarenal coagulation in rats. Thrombocytopenia and heparin cannot prevent intracortical fibrin formation. Thromb. Res. *34:* 87–92 (1984).
15 Losonczy, G.: Correlation between the early postischemic renal fibrin deposition and reduction of glomerular filtration rate in the rat. Effect of the defibrinating agent ARWIN. Acta physiol. hung. *66:* 183–188 (1985).
16 Sheehan, H.L.; Davis, J.C.: Renal ischemia with failed reflow. J. Path. Bact. *78:* 105–120 (1959).
17 Alkjaersig, N.; Fletcher, A.P.; Narayanan, M.; Robson, A.M.: Course and resolution of the coagulopathy in nephrotic children. Kidney int. *31:* 772–780 (1987).
18 Busch, C.; Rammer, L.: Quantitation of fibrin deposition and elimination in organs of rats injected with labelled fibrinogen by isolation of the labelled fibrin from water soluble tracer. Thromb. Diath. Haemorrh. *29:* 108–114 (1973).
19 Sheehan, H.L.: Renal morphology in preeclampsia. Kidney int. *18:* 241–252 (1980).
20 Losonczy, G.; Todd, H.; Palmer, D.C.; Hertelendy, F.: Prostaglandins, norepinephrine, angiotensin II and blood pressure changes induced by uteroplacental ischemia in rabbits. Clin. exp. Hypertens. *B5:* 271–294 (1986–87).
21 Thomas, L.; Good, R.A.: Studies on the generalized Shwartzman reaction. J. exp. Med. *96:* 605–624 (1952).
22 Ferris, T.F.: Postpartum renal insufficiency. Kidney int. *14:* 383–392 (1978).
23 Remuzzi, G.: HUS and TTP: variable expression of a single entity. Kidney int. *32:* 292–308 (1987).
24 Paller, M.S.; Hoidal, J.R.; Ferris, T.F.: Oxygen free radicals in ischemic acute renal failure in the rat. J. clin. Invest. *74:* 1156–1164 (1984).
25 Baud, L.; Nivez, M.P.; Chansel, D.; Ardaillou, R.: Stimulation by oxygen radicals of prostaglandin production by rat renal glomeruli. Kidney int. *20:* 332–339 (1981).
26 Patrono, C.; Dunn, M.J.: The clinical significance of inhibition of renal prostaglandin synthesis. Kidney int. *32:* 1–12 (1987).
27 Ham, E.A.; Egan, R.W.; Soderman, D.D.; Gale, P.H.; Kuehl, F.A., Jr.: Peroxidase-dependent deactivation of prostacyclin synthetase. J. biol. Chem. *254:* 2191–2194 (1979).
28 Hansson, R.; Gustafsson, B.; Jonsson, O.; Lundstam, S.; Pettersson, S.; Scherstén, T.; Waldenström, J.: Effect of xanthine oxidase inhibition on renal circulation after ischemia. Transplant. Proc. *14:* 51–60 (1982).

György Losonczy, MD, Semmelweis University Medical School, Department of Pathophysiology, POB 370, H-1445 Budapest (Hungary)

Contr. Nephrol., vol. 67, pp. 28–30 (Karger, Basel 1988)

Peroxidation of Lipids and Tissue Damage by Free Oxygen Radicals in the Pathogenesis of Inflammatory Kidney Diseases

Jurij E. Vel'tishchev

Institute of Paediatrics and Paediatric Surgery, Moscow, USSR

The importance of free oxygen radicals in biological processes was first noted by Emanuel in the USSR and Fridovich [3] in the USA several decades ago. It is now known that peroxidation is one of the basic reactions responsible for tissue renovation and for the ageing of living organisms. In recent years, increasing attention has been paid to the study of peroxidative processes in pathological conditions, and primarily in inflammation, when toxic oxygen radicals can destroy the lipid components of cell membranes, disrupt sulfhydryl bonds in proteins and enzymes, and also degrade DNA molecules.

The main source of free oxygen radical formation in the organism is different oxidative processes leading to the reduction of molecular oxygen and to the appearance of the superoxide anion radical. In many cells, and especially in phagocytic cells, NADP oxidation is the principal reaction initiating the peroxidative processes and halogenization needed to kill bacteria, etc. This reaction is reflected in an enhancement of the oxidative metabolism of the cells and in a respiratory burst.

We have studied some biochemical markers of lipid peroxidation in 125 children suffering from different inflammatory kidney diseases: obstructive pyelonephritis, interstitial nephritis and familial hereditary nephropathies [1, 2]. Nearly 30% of these children had different malformations of the kidney and the urinary tract, and also hypoplastic renal dysplasia. The methods used included measurement of the levels of conjugated dienes and malonic dialdehyde in the plasma and urine, the respiratory burst of the granulocytes, the nitroblue tetrazolium reduction test, and the activities of phospholipases A and C in blood. Special methods were used to detect substances containing acyl-nitrogen, Schiff bases and other products of peroxidative reactions. Experiments on animals (rats) were performed to reproduce interstitial nephritis with nephrocalcinosis by using toxic doses of cholecalciferol and ethylene glycol. In all cases of inflammatory kidney diseases, a marked activation of the lipid peroxidation was observed as well as increases of conjugated dienes and malonic dialdehyde in the plasma (2–3 times compared with the controls), an activation of the respiratory burst of granulocytes (by 50%), and enhancements of Schiff bases and 'polar' lipid excretion in the urine. In all cases, a lowering of the antioxidative power of the blood was noted. No significant differences were found in microbial and abacterial diseases. The influence of some microorganisms on the respiratory burst of the leukocytes was investigated. As a score, the reaction of cells to opsonized micrococci was selected (they bind on Fc- and C3b-receptors and produce the maximum rise in the oxidative metabolism of the granulocytes). Opsonized zymosan binding only on C3b-receptors had an activity two times lower. A very weak induction of the respiratory burst was found to be produced by staphylococci and *Escherichia coli.*

It is well known that pathogenic microorganisms have different and powerful systems of antiradical defence (superoxide dismutase, catalase, ion chelation, etc.). Insoluble crystals of calcium phosphate, oxalate and urate produced increases in the oxidative metabolisms of the granulocytes in vitro. For example, calcium phosphate dihydrate crystals caused a 1.8-fold rise in the oxidative metabolism of the granulocytes. In patients with interstitial nephritis and severe crystalluria, a close correlation was observed between the calcium oxalate excretion and the plasma level of malonic dialdehyde.

Thus, in microbial inflammation a surge of the superoxide anion radical can be connected with local and common immune reactions; in abacterial interstitial nephritis, one of the factors can be the deposition of insoluble crystals in the renal tissue.

Two varieties of N-acyl metabolites of membrane lipids were identified in the urine of patients and experimental animals. Their identification was carried out by preparative high-voltage electrophoresis and infrared spectroscopy of evaporated 'polar' lipids. The first group included different fatty acids amides; these were of the N-acyl-ethanol-amine type. The second group consisted of conjugates of fatty acids with substances containing different amino groups in which the nitrogen was protonated. Both types reflect severe membranodestructive processes in the kidney. However, they differ in biological activity. The first group, fatty acid amides, has no influence on chemotaxis and no cytolytic action on granulocytes. The second group, organic detergents, has very aggressive chemotactic and cytolytic properties. According to Del Maestro [4], protonated amine radicals are very active radical cations. Both species of N-acyl derivatives had activating effects on the oxidative metabolism of granulocytes. Other metabolites, such as 5-hydroxyeicosa-tetraenoic acid, were identified as products of lipoxygenase activation.

In most of the cases with inflammatory kidney disease, a decrease of the antioxidative power of the plasma was noted; this may serve as a basis for the introduction of antioxidative drugs (tocopherol, palmitoascorbate, mannitol and unithiol) in complex therapy. The use of new membrane-stabilizing agents combined with antioxidants seems more promising. We have obtained much better results by using some phosphonic acid derivatives (xidiphon and sodium etidronate) [1, 2].

These drugs depress the activity of the phospholipases, the excretion of 'polar' lipids and the aggregation of insoluble crystals, and they also chelate calcium and iron ions. Experimental interstitial nephritis and nephrocalcinosis caused by toxic doses of cholecalciferol were prevented with sodium etidronate.

Dimephosphone, a monophosphonic derivative, effectively enhanced the superoxide dismutase activity in the blood, abolished enzymuria and also modulated the immune response.

In conclusion, we would like to underline the importance of the activation of the peroxidative processes in inflammatory kidney disease. The investigation of such processes leads not only to new information on pathogenesis, but also to new diagnostic methods which may be useful in clinical nephrology, and to new ways for the more effective therapy of inflammatory kidney diseases.

References

1 Vel'tishchev, J.E.; Jurjeva, E.; Vosdvizenskaya, E.C.: Vopr. med. chimiii, Moscow. *2:* 2–9 (1987).

2 Ignatova, M.S.; Vel'tishchev, J.E.: Detskaya nephrologia, Leningrad 1982, p. 457.

3 Fridovich, J.: Oxygen radicals, hydrogen peroxide and oxygen toxicity; in Pryor, Free radicals in biology, vol. 1 (Academic Press, New York 1976).

4 Del Maestro, R.F.: An approach to free radicals in medicine and biology. Acta physiol. scand., suppl. 492, pp. 153–168 (1980).

Prof. Jurij Vel'tishchev, Institute of Paediatrics and Paediatric Surgery, Taldomskaya 2, Moscow 127 412 (USSR)

Contr. Nephrol., vol. 67, pp. 31–36 (Karger, Basel 1988)

A Kallikrein-Like Molecule and Plasma Vasoactivity in Minimal Change Disease[1]

Increased Turnover in Relapse versus Remission

W.W. Bakker[a], *J.F.W. Baller*[a], *W.H.J. van Luijk*[b]

[a] Department of Pathology, University of Groningen and
[b] Department of Pediatrics, University Hospital of Groningen, The Netherlands

The pathogenesis of increased glomerular permeability in minimal change nephrotic syndrome (MCNS) is often associated with immunologically mediated circulating factors [1–3]. The release of such factors, i.e. immunosuppressive [3] or vasoactive [1] lymphokines due to abnormal immune function is suggested to play a major role in the impairment of the glomerular filtration barrier in MCNS leading to albuminuria. Although it has been established now by several authors that mononuclear blood cells of subjects with MCNS in relapse are able to release a vasoactive factor after appropriate stimulation in vitro [4–6], a direct relationship of the action of such lymphokines and impairment of the glomerular filtration barrier is lacking [6].

Since loss of glomerular sialoglycoproteins, denoted as glomerular polyanion (GPA), has been described in MCNS during relapse [7], we focussed our attention on a vasoactive plasma factor present in normal human or rat plasma [8], which is able to affect GPA following incubation with kidney sections in vitro (fig. 1). This factor referred to as 100-KF (molecular weight in human plasma: 80–100 kdaltons) is also able to impair anionic sites within the endothelial part (lamina rara interna) of the GBM after perfusion of the rat kidney ex vivo [9] and displays kallikrein-like properties [8].

In the present communication we studied the amount of 100-KF in plasma samples from subjects with MCNS in relapse versus remission fol-

[1] This study was supported by grants from the Dutch Kidney Foundation.

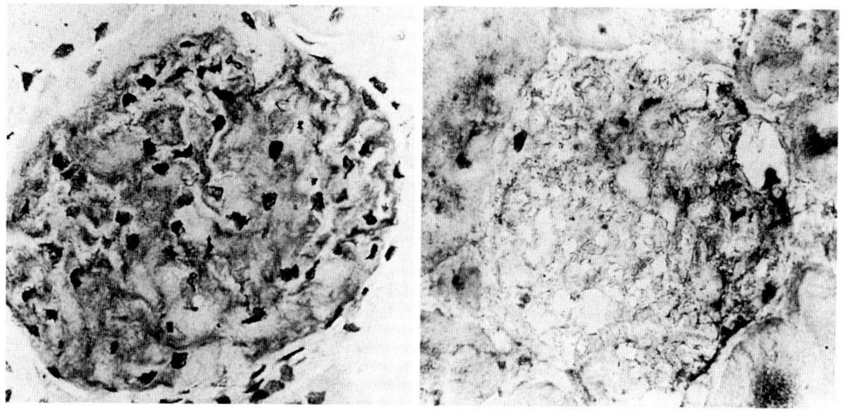

a b

Fig. 1. Glomeruli in sections of normal rat kidney stained for sialoprotein after incubation with: *(a)* control fraction (peak 1) or *(b)* 100-KF (5 mg protein/ml PBS). Colloidal iron. × 560.

lowing corticosteroid treatment using an antibody against 100-KF in the rocket electrophoresis assay according to Laurell [10]. The same samples were also studied for their vasoactive potention using a standard vascular permeability assay in the rat skin [6].

Materials and Methods

From 6 subjects aged 6–36 years with biopsy-proven MCNS (proteinuria 1.5–9.0 g/24 h; selectivity <17%, expressed as IgG clearance/albumin clearance × 100%) citrated blood was drawn and plasma samples were prepared according to standard methods and kept at −70 °C until use. From the same subjects in remission after 2–4 weeks after starting corticosteroid treatment (proteinuria <0.7 g/24 h) again plasma samples were prepared and stored at −70 °C.

Preparation of 100-KF from Normal Human Plasma

Isolation of 100-KF was carried out as described previously with minor modifications [8]. Dialysed pooled plasma (0.01 *M* Tris-HCl; 0.045 *M* NaCl pH 8.0) was run batchwise over a DE 52 ion-exchange column (4 °C) using subsequently 0.045, 0.08 and 1.0 *M* NaCl at constant pH 8.0. Concentrated fractions positive for vasoactivity and GPA-affecting capacity [8] were run over a Sephadex G150-SF column in PBS (pH 8.0), yielding two peaks which were assayed for their molecular weight using gradient polyacrylamide gel electrophoresis (PAGE) and again tested as described above (fig. 1). The positive fraction (molecular weight of 80–100 kdaltons) is referred to as 100-KF (fig. 2).

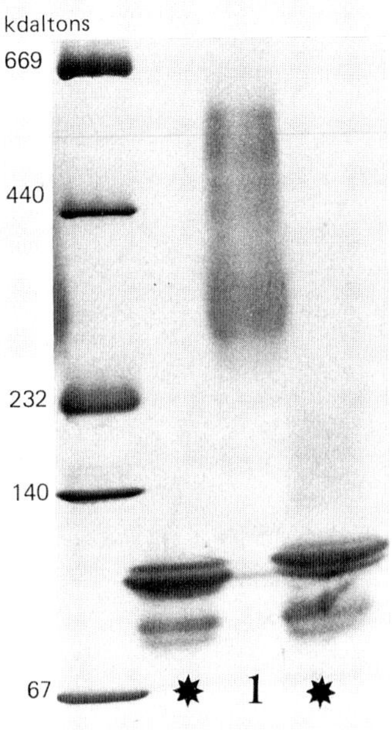

Fig. 2. Gradient PAGE of fraction eluted from Sephadex G150-SF columns. Fraction 2 (asterisk) contains 100-KF and control fraction 1 is seen in the middle lane (1). The left lane contains marker proteins (67 kdaltons bovine serum albumin, 140 kdaltons lactate dehydrogenase, 232 kdaltons catalase, 440 kdaltons ferritin, 669 kdaltons thyroglobulin).

Induction of Anti-100-KF IgG and Rocket Electrophoresis

Rabbits were immunized with 100-KF (5 mg protein/ml) in Freund's complete adjuvant according to standard procedures. IgG was isolated using column chromatography and tested for its monospecificity against 100-KF after thorough absorption to various insolubilized plasma fractions.

This antibody was used in the rocket electrophoresis assay according to Laurell [10] for evaluation of the 100-KF level in plasma samples from subjects in relapse versus remission. Also, plasma from normal donors was run in the Laurell assay together with serial dilutions of purified 100-KF.

Vascular Permeability Test (VPT)

Vascular permeability increasing activity of plasma samples was measured using a standard VPT in the rat as described elsewhere [6, 8]. Briefly: individual plasma samples containing 15% plasma in saline were given intradermally into the shaved backskins (70 μl per site) of ether-anesthetized Wistar rats (200 g) and a subsequent injection (250 μl) of Evans' blue (1% in saline) was given intravenously. Bradykinin (20 μg) and saline-injected sites in each animal served as positive or negative control spots.

After 60 min the animals were bled under ether anesthesia and the blueing areas, reflecting the extravasation of dye, were measured at the inner surface of the skin and expressed as a vascular permeability index (VPI):

$$\text{VPI} = \frac{\text{mean area of blueing around sample sites}}{\text{mean area of blueing around bradykinin sites}}.$$

3
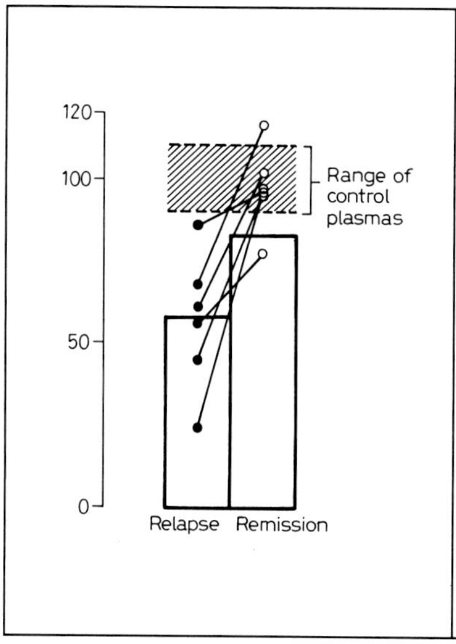

4
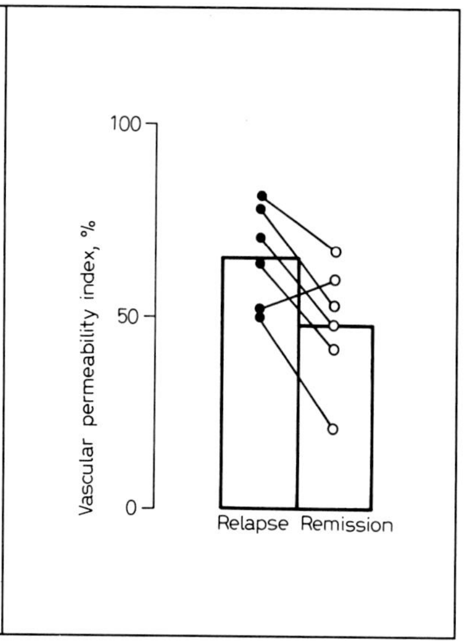

Fig. 3. Relative concentration of 100-KF in plasma from subjects with MCNS. Plasma samples from 5 of 6 subjects with MCNS in relapse showed increased vasoactivity as compared with plasma samples from those subjects in remission following corticosteroid treatment ($p \leq 0.05$ Wilcoxon).

Fig. 4. Vasoactivity of plasma from subjects with MCNS. Decreased level of 100-KF in plasma samples of subjects in relapse versus remission expressed as % of mean normal range ($p \leq 0.05$ Wilcoxon).

Results

The results show (fig. 3) that plasma samples from the majority of the nephrotic subjects before treatment showed enhanced vasoactivity as tested in vivo using our standard assay in the rat as compared to the samples taken from the subjects in remission. The mean of vascular permeability indexes in normal subjects (53%) is not significantly different from that of the remission samples.

In contrast, as can be seen from figure 4, the relative concentration of 100-KF in plasma from nephrotic subjects is significantly decreased as compared with samples from the same subjects in remission, the latter ones scoring within the same range of samples from normal donors ($n = 8$).

Discussion

The reduction of immunologically detectable amounts of 100-KF observed in plasma of the nephrotic subjects studied may be explained by either urinary loss and/or conversion of the enzyme due to inappropriate activation. Since the urine samples were not screened for 100-KF in this study the former possibility cannot be excluded completely. However, the concomitant increase of vasoactivity in the same plasma samples rather supports the alternative possibility, i.e. activation of 100-KF leading to increased consumption.

Activation of other factors of the kallikrein system, i.e. Hageman factor and prekallikrein, in MCNS during relapse has been also observed recently [11]. The mechanism whereby such activation occurs is uncertain and despite the fact that Hageman factor protein has been found in the urine of some subjects, simple excretory loss appears not to be an adequate explanation for the decreased plasma levels in MCNS [11, 12]. Although the possible direct relationship between increased plasma vasoactivity and decreased 100-KF protein in plasma of subjects in relapse remains to be proven, it is tempting to speculate that our kallikrein-like factor consists of a HMW kininogen-like molecule (MW $\approx$ 100 kdaltons) complexed with a kallikrein or prekallikrein-like molecule (MW $\approx$ 85 kdaltons) capable of releasing vasoactive kinins upon activation. Such activation may result in local impairment of GPA lining the endothelial sites, affection of anionic sites within the GBM as well as production of vasoactive molecules. Similarly, activation of Hageman factor within the nephrotic kidney has been suggested in MCNS [13] leading to depletion of prekallikrein associated with bradykinin release to mediate or enhance the increased glomerular vascular permeability in these patients.

As described by Kaller and Lee [13] for prekallikrein, our subjects showed similar restoration of the 100-KF plasma level in remission after corticosteroid treatment. The fact that corticosteroid treatment reduces plasma vasoactivity associated with normalisation of the 100-KF level supports the idea that the 100-KF level and plasma vasoactivity are directly related since corticosteroid treatment per se does not inhibit kinin release [14].

In conclusion, we observed, in patients with MCNS during relapse versus remission, an increased plasma vasoactivity associated with activation of a kallikrein-like plasma factor with the capacity of damaging the GBM resulting in proteinuria.

References

1 Lagrue, G.; Xheneumont, S.; Branallec, A.; Weil, B.: Lymphokines and nephrotic syndrome (Letter). Lancet *i:* 271–272 (1975).

2 Bakker, W.W.; Luyk, W.H.J. van; Hené, R.J.; Desmit, E.M.; Hem, G.K. van der; Vos, J.T.W.M.: Loss of glomerular polyanion in vitro induced by mononuclear blood cells from patients with minimal change nephrotic syndrome. Am. J. Nephrol. *6:* 107–111 (1986).

3 Schnaper, H.W.; Aune, T.M.: Steroid-sensitive mechanism of soluble immune response suppressor production in steroid-responsive nephrotic syndrome. J. clin. Invest. *79:* 257–264 (1987).

4 Thomizawa, A.; Suzuki, S.; Oguri, M.; Kuroume, T.: Studies of T lymphocyte function and inhibitory factors in minimal change nephrotic syndrome. Nephron *24:* 179–182 (1978).

5 Boulton Jones, J.M.; Simpson, S.: Immunological studies of minimal change glomerulopathy. Br. med. J. *280:* 291–293 (1980).

6 Bakker, W.W.; Beukhof, J.R.; Luyk, W.H.J. van; Hem, G.K. van der: Vascular permeability increasing factor in IgA nephropathy. Clin. Nephrol. *8:* 165–167 (1982).

7 Blau, E.B.; Haas, J.E.: Glomerular sialic acid and proteinuria in human renal disease. Lab. Invest. *28:* 477–482 (1973).

8 Bakker, W.W.; Roskam, G.; Hardonk, M.D.; Vos, J.T.W.M.; Bleumink, E.: The glomerular polyanion (GPA) of the rat kidney. III. Further characterization of a vasoactive serum factor which reduces GPA. Br. J. exp. Path. *66:* 47–55 (1985).

9 Bakker, W.W.; Donga, J.; Koudstaal, J.; Baller, J.F.W.: Alteration of glomerular basement membrane (GBM) anionic sites of the rat kidney by a vasoactive kallikrein-like factor isolated from human plasma (Abstract). Xth Int. Congr. Nephrol., London 1987, p. 372.

10 Laurell, C.B.: Electroimmunoassay. Scand. J. clin. Lab. Invest. *29:* 21–37 (1972).

11 Hruby, M.A.; Honig, G.R.; Shapira, E.: Immunoquantitation of Hageman factor in urine and plasma of children with nephrotic syndrome. J. Lab. clin. Med. *96:* 501–510 (1980).

12 Alkjaersig, N.: The coagulopathy of nephrotic syndrome; in Brodehl, Ehrich, Pediatric nephrology, pp. 257–258 (Springer, Heidelberg 1984).

13 Kaller, R.J.; Lee, S.K.: A study of the plasma kinin generating system in children with the minimal lesion idiopathic nephrotic syndrome. Pediat. Res. *9:* 705–709 (1975).

14 Nakakara, M.: Plasma hippuryl-*L*-lysine hydrolase in tourniquet shock. Experientia *29:* 999–1000 (1973).

W.W. Bakker, MD, Department of Pathology, Oostersingel 63,
NL–9713 EZ Groningen (The Netherlands)

Contr. Nephrol., vol. 67, pp. 37–43 (Karger, Basel 1988)

Disturbances of Lipoprotein and PGI_2 Metabolism in IgA Nephropathy and Henoch-Schönlein Purpura

S. Túri[b], *J. Nagy*[c], *I. Haszon*[a], *Z. Havass*[a], *M. Németh*[b], *Cs. Bereczki*[a]

[a]Department of Pediatrics, and
[b]Department of Clinical Chemistry, University Medical School, Szeged;
[c]2nd Department of Medicine, University Medical School, Pécs, Hungary

Current consensus classifies IgA nephropathy as an immune-complex (IC) glomerulonephritis in which most of the antibody deposited is IgA. The overlap in clinical, pathologic and immunologic features of primary IgA nephropathy and Henoch-Schönlein purpura has been widely appreciated. The chief differences are multiorgan system involvement, more frequent glomerular tuft necrosis, increases in circulating IC-containing IgA, and the likelihood of extrarenal deposits of IgA and C_3 in the latter [1], but the differences appear to be more quantitative than qualitative.

During this inflammatory process, a variety of stimuli can induce the liberation of membrane phospholipids and arachidonic acid (ARA). Data are available concerning the possible roles of vascular prostaglandins and thromboxane (TxA_2) in the pathogenesis of vasculopathies. PGI_2 is an unstable substance with potent antiaggregatory and vasodilatory activity. It is synthesized from ARA by the vessel wall cyclo-oxygenase system [2]. The other unstable endoperoxide with vasoconstricting and platelet-aggregating activity [3] is TxA_2, derived from ARA predominantly by the platelet cyclo-oxygenase system [4]. Remuzzi and co-workers first reported a deficient PGI_2 production in haemolytic uraemic syndrome, thrombotic thrombocytopenic purpura [5], and severe pre-eclampsia [6], while increased TxA_2 synthesis by platelets has been shown in Kawasaki disease [7]. In Henoch-Schönlein purpura a lower concentration of plasma PGI_2

metabolite (6-keto-$PGF_1\alpha$) was observed, with a decreased ability of the plasma to support vascular PGI_2 production in vitro [8].

We report here our recent findings on the various aspects of PGI_2 metabolism in IgA glomerulonephritis and Henoch-Schönlein purpura.

Patients and Methods

Patients

Forty-five patients with IgA glomerulonephritis (aged 35.2±12.9 years, mean ± SD), 18 with Henoch-Schönlein purpura (aged 6.8 ± 2.7 years), 8 of them combined with nephrotic syndrome, and 17 paediatric and 24 adult controls (aged 5.6 ± 2.8 and 33.8 ± 13.7 years, respectively), matched for age and sex, were studied. Serum creatinine was elevated in 9 (201.3 ± 87.1 µmol/l) and normal in 36 (92.4 ± 14.2 µmol/l) patients with IgA nephropathy. Eighteen had a urinary protein excretion of more than 0.5 g/24 h (1.48 ± 1.02 g/24 h). Average time of follow-up for these patients was 10.4 ± 7.3 years. All patients with Henoch-Schönlein vasculitis had the classical purpuric rash, and in 15 cases the skin lesions were associated with joint manifestations. In 11 patients, gastrointestinal symptoms were observed in 7 patients haematuria (> 10 red blood cells/µl), and in 8 patients proteinuria (> 80 mg/h/m^2) was associated. Altogether 9 patients had abnormal urinalysis (the creatinine clearance in these cases was 56.7 ± 11.08 ml/min/1.73 m^2). Patients without urinary symptoms had a significantly higher creatinine clearance (98.8 ± 20.9 ml/min/1.73 m^2).

Methods

After informed consent had been obtained, 10 ml of venous blood was taken from the controls and from the patients with IgA nephropathy and Henoch-Schönlein purpura 7–14 days after the occurrence of the rash. Blood for estimation of the ability of the plasma to support PGI_2-like activity (PSA) was anticoagulated with 3.2% trisodium citrate at a ratio of 9:1. Samples for total, high-density and low-density lipoprotein cholesterol (HDL and LDL cholesterol) examination and gas-liquid chromatography were collected with 3.8% w/v trisodium citrate in a 9:1 volume (blood:Na_3 citrate).

No patient or control had been taking any drug known to alter the prostaglandin metabolism for at least 2 weeks before the study.

For the measurement of the plasma ability to support PGI_2 production (PSA) from human umbilical arterial rings, the method of Moncada et al. [9], was used, which is based on the platelet antiaggregatory effect of the ring's supernatant [8, 10]. To detect the inhibitory activity in the test plasma against PGI_2-like effect (PSAI), a modified method of Levin et al. [11] was used, which we have reported elsewhere [12]. Plasma fatty acid component measurements were carried out by gas-liquid chromatography [13]. Plasma total cholesterol and HDL cholesterol were determined with a Gödecke kit [14, 15]. LDL cholesterol studies were performed with a Böhringer kit [16].

All platelet aggregation, gas-liquid chromatography and cholesterol studies were carried out in duplicate. Statistical analysis was performed with Student's t test and the Spearman rank correlation test.

Results

Ability of Plasma to Support PGI_2-Like Activity

This was reduced in the patients with IgA nephropathy who were studied. The plasma PSA results are given in table I. In 12 cases, PSA was undetectable. The difference between the results for the IgA nephropathy group and the adult control group was significant ($p < 0.05$). The plasma PSA was also reduced in the group of patients with Henoch-Schönlein purpura. In 6 cases PSA was undetectable. There was a significant difference between the results for patients with Henoch-Schönlein purpura only and those with Henoch-Schönlein purpura combined with nephrotic syndrome ($p < 0.05$). The difference between the results for the Henoch-Schönlein purpura group (18 patients) and the paediatric control group was significant ($p < 0.05$). We found no significant difference between the results for the Henoch-Schönlein purpura group and the patients with IgA nephropathy (table I).

Detection of Inhibition against PGI_2-Like Activity

Plasma samples from 30 patients with IgA nephropathy, and 10 with Henoch-Schönlein purpura (8 of them combined with nephrotic syndrome), in which the PSA was low or absent, were further investigated to detect PSAI. Less PSA was produced by the unexhausted umbilical arterial rings when they were incubated with platelet-poor plasma from 23 patients with IgA nephropathy and all 10 patients with Henoch-Schönlein purpura.

Concentrations of Total, HDL and LDL Cholesterol in Plasma

These investigations were carried out in 18 cases from each group. The plasma concentrations of total cholesterol and LDL cholesterol were sig-

Table I. Ability of plasma to support PGI_2-like activity from vascular endothelium

	n	PSA, %	
IgA nephropathy	45	27.6 ± 22.3	$p < 0.05$
Henoch-Schönlein purpura (HSP)	18	20.0 ± 19.1	$p < 0.05$
HSP	10	31.1 ± 22.1	
HSP + nephrotic syndrome	8	14.3 ± 11.7	$p < 0.01$
Paediatric control	17	54.4 ± 7.5	
Adult control	24	61.3 ± 21.6	

nificantly higher in patients with IgA nephropathy than in the controls ($p < 0.01$). The HDL cholesterol level in the plasma was significantly lower in the patients with IgA nephropathy compared with the results of the controls ($p < 0.01$). In the plasma from patients with Henoch-Schönlein purpura the HDL cholesterol concentration was significantly lower than in the controls ($p < 0.01$). The total cholesterol and LDL cholesterol levels were increased in the plasma from patients with Henoch-Schönlein purpura, but the difference was not significant relative to the controls. The ratio between plasma LDL and HDL was significantly higher in IgA nephropathy ($p < 0.01$) and Henoch-Schönlein purpura ($p < 0.05$) than in the controls. A negative correlation was found between the plasma LDL/HDL ratio and the PSA level in IgA nephropathy ($r = -0.65$, $p < 0.05$) and in Henoch-Schönlein purpura ($r = -0.57$, $p < 0.05$) (table II). Patients with PSAI in their plasma exhibited a high concentration of LDL cholesterol and/or a low level of HDL cholesterol.

Measurements of Plasma Fatty Acid Components by Gas-Liquid Chromatography

The detailed results of the determination of plasma fatty acid components belonging to phospholipids, triglycerides and free fatty acids will be reported elsewhere. We present here the values of the prostaglandin precursors ARA and dihomogamma-linolenic acid. The percentage of ARA in the fatty acid components of the plasma phospholipids was 8.54 ± 2.37% (mean ± SD) in patients with Henoch-Schönlein purpura, 8.08 ± 3.07% in IgA nephropathy and 7.47 ± 2.65% in the control group. The differences between the groups of patients and the controls were not significant. The percentage of dihomogamma-linolenic acid in the fatty acid components of the plasma phospholipids was 2.40 ± 1.03% in the plasma of patients with Henoch-Schönlein purpura, 2.28 ± 1.36% in patients with

Table II. Plasma cholesterol levels in Henoch-Schönlein purpura (HSP) and IgA nephropathy (IgA-gn)

	n	Total cholesterol mmol/l		LDL cholesterol mmol/l		HDL cholesterol mmol/l		LDL/HDL cholesterol	
HSP	18	3.94±0.75		2.53±0.51		0.86±0.27	<0.01	3.16±0.80	<0.05
IgA-gn	18	5.11±1.00	<0.01	3.56±0.40	<0.01	0.80±0.17	<0.01	4.59±0.97	<0.01
Control	18	3.61±0.55		2.19±0.31		1.28±0.25		2.10±0.49	

IgA nephropathy and 1.79 ± 0.74% in the controls. There was a significant difference between the values for patients with Henoch-Schönlein purpura and the controls ($p < 0.05$).

Discussion

The results of the present study provide confirmatory data on the deficient vascular PGI_2 production on the action of plasma (PSA) from patients with Henoch-Schönlein purpura. These findings were very similar to those obtained with the plasma from adult patients with IgA nephropathy, which suggests a common feature in the pathophysiology of the two diseases. All 10 patients with Henoch-Schönlein purpura who were studied and 23 of the 30 cases with IgA nephropathy showed the presence of PSAI in their plasma. These patients had a low or zero level of PSA which was in accordance with our previous study [8]. The presence of PSAI may mean inhibitory activity against PGI_2 production [11], inhibitory activity against the PGI_2 effect [8], or increased PGI_2 degradation [17]. We demonstrated with gas-liquid chromatography that the levels of ARA and the ARA precursor were not lower in the plasma of patients than in the controls. Therefore, the decreased ability of the plasma to support vascular PGI_2 synthesis in Henoch-Schönlein purpura and IgA nephropathy cannot be explained by a low level of PG precursors.

Recent investigations have shown that a high concentration of LDL can inhibit PGI_2 biosynthesis [18–21], while HDL stimulates it [22, 23] under in vitro conditions, with significant differences between male and female donors. In our study the plasma concentrations of total and LDL cholesterol were significantly higher in patients with IgA nephropathy, while that of HDL cholesterol was lower in both the IgA nephropathy and Henoch-Schönlein purpura groups than in the controls. We found a negative correlation between the plasma LDL/HDL ratio and the PSA levels in IgA nephropathy and Henoch-Schönlein purpura. Patients with PSAI in their plasma exhibited a high concentration of LDL cholesterol and/or a low level of HDL cholesterol. We found no difference between the results for male and female patients. These findings point to an inhibitory effect of LDL and a protective role of HDL in the PGI_2 synthesis of patients with IgA nephropathy and Henoch-Schönlein purpura.

It is very likely that the decreased vascular PGI_2 formation is a secondary phenomenon in the pathogenesis of vasculitis. The primary event is the vascular damage caused by IC deposition. This, combined with the changes in the plasma lipoprotein constituents and in the ratio of the lipoproteins, may influence the vascular PGI_2 production. These findings are probably not specific for IgA nephropathy or Henoch-Schönlein purpura. We recently found a positive correlation between the PSA and HDL cholesterol values and a negative correlation between the PSA and LDL cholesterol levels in the plasma from patients with diabetic microangiopathy [24].

Apart from the oxidation products of the cyclo-oxygenase pathway, hydroxy fatty acids (hydroxyeicosatetraenoic acids) and leukotrienes are synthesized via the lipoxygenase pathway, which can also influence the pathophysiologic events in glomerulonephritis and other vasculitis syndromes [25]. Therefore, the deficient PGI_2 production is not the only factor responsible for glomerular disease. Nevertheless, the absence of PGI_2 could play an important role in the progression of the angiopathy.

References

1 Coppo, R.; Basolo, B.; Martina, G.; Rollino, C.; De Marchi, M.; Ciacchino, F.; Mazzucco, G.; Messina, M.; Piccoli, G.: Circulating immune complexes containing IgA, IgG and IgM in patients with primary IgA nephropathy and with Henoch-Schönlein nephritis. Correlation with clinical and histologic signs of activity. Clin. Nephrol. *18:* 230–239 (1982).

2 Moncada, S.; Gryglewski, R.; Bunting, S.; Vane, J.R.: An enzyme isolated from arteries transforms prostaglandin endoperoxides to an unstable substance that inhibits platelet aggregation. Nature, Lond. *263:* 633–635 (1976).

3 Packham, M.A.; Mustard, J.F.: Pharmacology of platelet-affecting drugs. Circulation *62:* suppl. V, pp. 26–32 (1980).

4 Hamberg, M.; Svenson, J.; Samuelson, B.: Thromboxanes: a new group of biologically active compounds derived from prostaglandin endoperoxides. Proc. natn. Acad. Sci. USA *72:* 2994–2998 (1975).

5 Remuzzi, G.; Marchesi, D.; Mecca, G.; Misiani, R.; Livio, M.; De Gaetano, G.: Haemolytic-uraemic syndrome: deficiency of plasma factor(s) regulating prostaglandin activity? Lancet *ii:* 871–872 (1978).

6 Remuzzi, G.; Marchesi, D.; Zoja, C.; Muratore, D.; Mecca, G.; Misiani, R.; Rossi, E.; Barbato, M.; Capetta, P.; Donati, M.B.; De Gaetano, G.: Reduced umbilical and placental vascular prostacyclin in severe pre-eclampsia. Prostaglandins *20:* 105–110 (1980).

7 Hidaka, T.; Nakano, M.; Ueta, T.; Komatsu, Y.; Yamanoto, M.: Increased synthesis of thromboxane A_2 by platelets from patients with Kawasaki disease. J. Pediat. *102:* 94–96 (1983).

8 Túri, S.; Belch, J.J.F.; Beattie, T.J.; Forbes, C.D.: Abnormalities of vascular prostaglandins in Henoch-Schönlein purpura. Archs Dis. Childh. *61:* 173–177 (1986).

9 Moncada, S.; Higgs, E.A.; Vane, J.R.: Human arterial and venous tissue generate prostacyclin (prostaglandin X) a potent inhibitor of platelet aggregation. Lancet *i:* 18–20 (1977).
10 Born, G.V.R.: Aggregation of platelets by ADP and its reversal. Nature, Lond. *194:* 927–929 (1962).
11 Levin, M.; Elkon, K.B.; Nokes, T.J.C.; Buckle, A.M.; Dillon, M. J.; Hardisty, R.M.; Barratt, T.M.: Inhibitor or prostacyclin production in sporadic hemolytic uremic syndrome. Archs Dis. Childh. *58:* 703–708 (1983).
12 Túri, S.; Beattie, T.J.; Belch, J.J.F.; Murphy, A.V.: Disturbances of prostacyclin metabolism in children with hemolytic-uremic syndrome and in first degree relatives. Clin. Nephrol. *25:* 193–198 (1986).
13 Mailanen, T.; Nikkari, T.: The effect of storage on the fatty acid composition of human serum. Clinica chim. Acta *114:* 111–116 (1981).
14 Allain, C.C.; Poon, I.S.; Chan, C.S.G.; Richmond, W.; Fu, P.C.: Enzymatic determination of total serum cholesterol. Clin. Chem. *20:* 470–475 (1974).
15 Lopes-Virella, M.E.; Store, P.; Ellis, S.; Colwell, I.A.: Cholesterol determination in high density lipoproteins separated by three methods. Clin. Chem. *23:* 882–884 (1977).
16 Assmann, G.; Jabs, H.V.; Kohnert, V.; Nolte, W.; Schriewer, H.: LDL-cholesterol determination in blood serum following precipitation of LDL with polyvinylsulphate. Clinica chim. Acta *140:* 77–83 (1984).
17 Chen, Y.C.; McLeod, B.; Hall, E.R.; Wu, K.K.: Accelerated prostacyclin degradation in thrombotic thrombocytopenic purpura. Lancet *ii:* 268–269 (1981).
18 Nordøy, A.; Svensson, B.; Wiehler, D.; Hoak, I.C.: Lipoproteins and the inhibitory effects of human endothelial cells on platelet function. Circulation Res. *43:* 527–534 (1978).
19 Nordøy, A.: Lipids and thrombogenesis. Ann. clin. Res. *13:* 50–61 (1981).
20 Szczeklik, A.; Cryglewski, R.J.: Low density lipoproteins (LDL) are carriers for lipid peroxides and inhibit prostacyclin (PGI_2) biosynthesis in arteries. Artery *7:* 488–495 (1980).
21 Beitz, J.; Förster, W.: Differential influence of lipoproteins isolated from women and men on the activity of the PGI_2 synthetase activity. Prostaglandins Leukotrienes Med. *6:* 515–518 (1981).
22 Fleisher, L.N.; Tall, A.R.; Witte, L.D.; Miller, R.W.; Cannon, P.J.: Stimulation of arterial endothelial cell prostacyclin synthesis by high density lipoproteins. J. biol. Chem. *257:* 6653–6655 (1982).
23 Giessler, C.H.; Beitz, J.; Mentz, P.; Förster, W.: The influence of lipoproteins (LDL and HDL) on PGI_2 formation by isolated aortic preparations of rabbits. Prostaglandins Leukotrienes Med. *22:* 221–234 (1986).
24 Túri, S.; Magyari, M.; Németh, M.; Bereczki, Cs.: Plasma factors influencing prostacyclin-like activity in patients with diabetic microangiopathy. Prostaglandins Leukotrienes Med. (in press).
25 Lianos, E.A.: Biosynthesis and role of arachidonic acid metabolites in glomerulonephritis. Nephron *37:* 73–77 (1984).

S. Túri, MD, Department of Paediatrics, University Medical School,
14–15 Korányi Str., H-6725 Szeged (Hungary)

Contr. Nephrol., vol. 67, pp. 44–47 (Karger, Basel 1988)

Arginine Vasopressin in Children with Moderately Impaired Renal Function

Plasma Concentration and Its Relationship to Urinary Clearance[1]

M. Bald, W. Rascher, K. Schärer

Division of Pediatric Nephrology, Children's Hospital,
University of Heidelberg, FRG

The antidiuretic hormone arginine vasopressin (AVP) is regulated primarily by the osmotic pressure in plasma and influenced by nonosmotic mechanisms such as intravascular volume or arterial blood pressure [6, 11]. Measurement of AVP in plasma directly reflects the biological activity of the hormone at the moment of blood withdrawal, whereas urinary excretion of AVP is considered to be a measure of its integrated activity over a longer period of time. However, the validity of this assumption remains to be confirmed, at least in the status of impaired renal function, because AVP is metabolized in the kidney [12]: the hormone is filtered in the glomerulus, reabsorbed and/or inactivated in the proximal tubule and secreted in the distal tubule. In addition, inactivation of AVP takes place in the peritubular circulation [5], probably mediated by receptor binding [13].

Recent studies from our laboratory have demonstrated elevated plasma concentrations of AVP in children with chronic renal failure [8]. To evaluate whether this is due to renal excretion of AVP, we examined the relationship between renal blood flow, glomerular filtration rate, actual plasma concentration and urinary clearance of AVP in children with moderately impaired renal function.

[1] This study was supported by the Deutsche Forschungsgemeinschaft (Ra 326/1-2).

Methods

Twenty-two children (11 males, 11 females, median age 11.9 years, range 8.9–17.9 years) were studied during a routine renal function test for progression of renal disease. Measurement of para-aminohippuric acid (PAH) and inulin clearances were reported elsewhere [4]. AVP in plasma and urine was measured by radioimmunoassay [7], whereby urine was directly assayed without prior extraction. Plasma and urine osmolalities were measured by a vapour pressure osmometer (Wescor, Inc., Logan, Utah). Serum creatinine was measured by a routine laboratory method and clearances were calculated using standard formulas.

Results and Discussion

The results of the clearance studies are shown in table I. In 41% of our patients, UC_{AVP} was greater than C_{IN} and also mean UC_{AVP} exceeded mean C_{IN}, but this difference was not statistically significant.

This is in contrast to the results of Ardaillou et al. [1] in healthy volunteers of patients with renal failure, in which UC_{AVP} was less than 10% of C_{IN}. However, our data fit those found by stop-flow analysis in the dog [3]. Therefore, we conclude that AVP is not only filtrated in the glomerulus, but also secreted in the distal tubule.

Although C_{PAH} and C_{IN} correlated strongly ($r = 0.90$, $p < 0.01$), there was no significant relationship between either C_{PAH} or C_{IN} and UC_{AVP}, but we found a good positive correlation between FE_{AVP} (= UC_{AVP}/C_{IN}) and UC_{AVP} ($r = 0.78$, $p < 0.01$). This indicates that the urinary clearance of AVP is independent of renal blood flow or glomerular filtration rate in children with moderately impaired renal function.

Table I. Plasma and urine concentrations of AVP, serum creatinine and clearance values in 22 children with moderately impaired renal function ($\bar{x} \pm SEM$)

Plasma AVP (P_{AVP}), pg/ml	3.2 ± 0.4
Urine AVP (U_{AVP}), pg/ml	54.8 ± 11.7
Serum creatine (S_{CR}), mg/dl	1.0 ± 0.2
PAH clearance (C_{PAH}), ml/min × 1.73 m^2	373.9 ± 43.3
Inulin clearance (C_{IN}), ml/min × 1.73 m^2	68.1 ± 7.1
Osmolar clearance (C_{OSM}), ml/min × 1.73 m^2	2.3 ± 0.2
Free water clearance (C_{H_2O}), ml/min × 1.73 m^2	0.9 ± 0.3
Urinary AVP clearance (UC_{AVP}), ml/min × 1.73 m^2	73.5 ± 25.4

C_{OSM} correlated positively with UC_{AVP} ($r = 0.70$, $p < 0.01$), as already reported by Robertson [10], but C_{H_2O} was independent from UC_{AVP}.

A positive correlation has been reported between the concentration of AVP in plasma and urine in healthy volunteers [2] as well as in newborn infants [9]. In our study, we could not confirm these results, because we found no relationship between P_{AVP} and urine AVP concentration. It should be mentioned that the urinary excretion rate of AVP depends on the state of hydration [10] and on renal function [1]. Therefore, a direct relation can only be achieved under controlled clinical conditions.

Finally, we found a strong negative relationship between P_{AVP} and log UC_{AVP} ($r = -0.71$, $p < 0.01$), indicating that increased plasma levels of AVP are associated with reduced urinary AVP clearance in children with impaired renal function. It is suggested that there is an active, probably receptor-mediated, mechanism of urinary AVP excretion which can be saturated by high plasma levels of AVP in renal failure.

References

1 Ardaillou, R.; Pruszczynski, W.; Benmansour, M.: Secretion and catabolism of antidiuretic hormone in renal failure. Contr. Nephrol., vol. 50, pp. 46–53 (Karger, Basel 1986).

2 Khokar, A.M.; Slater, J.D.H.; Forsling, M.L.; Ramage, C.M.: The physiological significance of urinary vasopressin and its relationship to plasma levels in man. Clin. Sci. mol. Med. *51:* 14p–15p (1976).

3 Kimura, T.; Share, L.: Characterization of the renal handling of vasopressin in the dog by stop-flow analysis. Endocrinology *109:* 2089–2094 (1981).

4 Manz, F.; Alatas, H.; Kochen, W.; Lutz, P.; Rebien, W.; Schärer, K.: Determination of glomerular function in advanced renal failure. Archs Dis. Childh. *52:* 721–724 (1977).

5 Rabkin, R.; Share, S.; Payne, P.A.; Young, J.; Crofton, J.: The handling of immunoreactive vasopressin by the isolated perfused rat kidney. J. clin. Invest. *63:* 6–13 (1979).

6 Rascher, W.; Lang, R.E.; Unger, T.: Vasopressin, cardiovascular regulation, and hypertension; in Ganten, Pfaff, Current topics in neuroendocrinology, pp. 101–136 (Springer, Berlin 1984).

7 Rascher, W.; Lang, R.E,; Unger, T.; Ganten, D.; Gross, F.: Vasopressin in brain of spontaneously hypertensive rats. Am. J. Physiol. H *242:* H496–H499 (1982).

8 Rauh, W.; Hund, E.; Sohl, G.; Rascher, W.; Mehls, O.; Schärer, K.: Vasoactive hormones in children with chronic renal failure. Kidney int. *24:* suppl. 15, pp. S27–S33 (1983)

9 Rees, L.; Brook, C.G.D.; Forsling, M.L.: Continuous urine collection in the study of vasopressin in the newborn. Hormone Res. *17:* 134–140 (1983).

10 Robertson, G.L.: The regulation of vasopressin function in health and disease. Rec. Prog. horm. Res. *33:* 333–385 (1977).
11 Schrier, R.W.; Berl, T.; Anderson, R.J.: Osmotic and nonosmotic control of vasopressin release. Am. J. Physiol. F *236:* F321–F332 (1979).
12 Share, L.; Kimura, T.; Matsui, K.; Shade, R.E.; Crofton, J.T.: Metabolism of vasopressin. Fed. Proc. *44:* 59–61 (1985).
13 Weitzman, R.E.; Fisher, D.A.: Arginine vasopressin metabolism in dogs. I. Evidence for a receptor-mediated mechanism. Am. J. Physiol. E. *236:* E591–E597 (1978).

M. Bald, MD, Universitätsklinik für Kinder- und Jugendmedizin (GHS), Hufelandstrasse 55, D–4300 Essen (FRG)

Contr. Nephrol., vol. 67, pp. 48–52 (Karger, Basel 1988)

Release of Human Natriuretic Peptide in Kidney Transplantation

G. Offner, P.F. Hoyer, R. Galaske, B. Oemar, H. Jueppner, C. Kroeger
Department of Paediatrics, University Medical School, Hannover, FRG

Volume loading in renal transplantation is an accepted factor for the induction of primary graft function. In our center volume loading was performed in about 100 pediatric kidney transplantations since 1979 which resulted in a primary function rate of 85% [1]. One major factor for this observation could be the release of atrial peptides by the distention of the right atrium, which evokes natriuresis and diuresis [2]. We therefore studied the plasma concentrations of human natriuretic peptide (h-ANP) in 14 patients who underwent renal transplantation in the last 10 months.

Patients and Methods

The biological data of the 14 patients are listed in table I. The mean age was 11.9 years with a range between 2.6 and 22.1 years. The sex ratio was 8 females to 6 males. The source of kidney grafts was in 1 case the living father and in 13 from dead donors. Eleven children had a first graft, 2 a second and 1 a third. Dialysis methods before transplantation were CAPD in 6 children and hemodialysis in 5 children, while 3 children could be transplanted without prior dialysis.

Immunosuppression was started with ciclosporin, 500 mg/m^2 p.o. 6 h preoperatively, and continued at 250 mg/m^2 after 12 h. Prednisolone was started at 300 mg/m^2 intraoperatively and continued at 150 mg/m^2 after 12 h. Mannitol was given intraoperatively 1 g/kg before declamping the anastomosis. Further volume loading was performed with albumin at a dosage of 4 g/kg/24 h in order to keep the central venous pressure at 7 mm Hg. Ringer's lactate was started at 100 ml/m^2/h and continued corresponding to diuresis. Furosemide was started at 10 mg/kg during transplantation and continued at 10 mg/kg during the following 24 h. Dopamine was given at a renal perfusion dose of 5 mg/m^2/h. Primary function of the graft was defined by a diuresis of 100 ml/m^2/h 6 h after transplantation and a decrease of serum creatinine within 24 h.

Table I. Biological data of 14 patients

Mean age, years (range)		11.9 (2.6–22.1)
Sex, F/M		8/6
Graft	LRD/CD	1/13
	1/2/3	11/2/1
CAPD/HD/without dialysis		6/5/3

LRD = Living renal donor; CD = cadaveric donor; HD = hemodialysis.

Table II. Results: volume loading

		Mean	Range
Performed volume loading $ml/m^2/24$ h	mannitol (20% solution)	177	107–391
	albumin (20% solution)	341	132–696
	Ringer's lactate	7,584	3,264–13,920
CVP, mm Hg	minimum	4.1	1.4–7.7
	maximum	7.2	3.2–9.9

ANP was measured in unextracted plasma samples using an antisheep radioimmunoassay from the IBL Company (Hamburg, FRG). Four 2-ml blood samples were taken into cooled syringes containing EDTA (0.05 m*M*) and aprotinine (10,000 U) preoperatively, during transplantation when the graft perfusion was started, and 6 and 24 h after transplantation.

Results

The 14 patients received during the first 24 h after kidney transplantation a mean dose of mannitol as a 20% solution of 177 ml/m^2, of 20% albumin of 341 ml/m^2 and of Ringer's lactate of 7,584 ml/m^2. The mean central venous pressure was between 4.1 and 7.2 mm Hg (table II).

The outcome of graft function is summarized in table III. Surgical interventions within the first 24 h were necessary in 2 cases, due to a twisted renal vein in 1 case and severe bleeding with graft loss in the second case. Thirteen grafts were functioning thereafter, which means 93% graft survival. Nine grafts from 12 without technical complications had a primary function which was 75%, 20% higher than reported from other

Table III. Graft function (n = 14)

Surgical complications		2	14%
Functioning graft		13	93%
Primary function		9	75%
Mean diuresis, ml/m²/24 h (range)		5,840	(60–13,140)
Mean serum creatinine (range)	before transplant	697	(333–1,098)
	24 after transplant	454	(144–668)

centers [3]. The mean diuresis of all 14 patients was 5,840 ml during the first 24 h. The mean serum creatinine decreased from 697 to 454 μmol/l.

The mean plasma concentrations of h-ANP were already increased before transplantation with a mean value of 0.30 vs. 0.10 ng/ml in a normal control group of healthy students [4]. Control values of healthy children were not known. After volume loading there was a significant increase of h-ANP to 0.46 ng/ml intraoperatively with a further increase after 6 and 24 h on 0.70 and 0.64 ng/ml, respectively (fig. 1). The individual plasma concentrations of h-ANP increased after renal transplantation in all 14 patients except 1 who already had a high ANP level of 0.4 ng/ml preoperatively (fig. 2). The difference was statistically significant at $p < 0.01$. It is notable that the lower preoperative ANP levels belonged to the children on CAPD.

There was no correlation between the increase of h-ANP and graft function. This is interpreted as evidence that other factors of graft damage such as cold ischemia time, histocompatibility and surgical complications influenced the graft function to a major extent, which cannot be overcome by ANP.

Conclusions

The plasma concentrations of ANP were elevated in children with end-stage renal failure before kidney transplantation, according to Rascher et al. [5]. Volume loading with mannitol, albumin and Ringer's lactate during kidney transplantation led to a further significant increase of ANP which might be one factor contributing to early diuresis. The effects of ANP and volume loading seem to be restricted to grafts without severe damage.

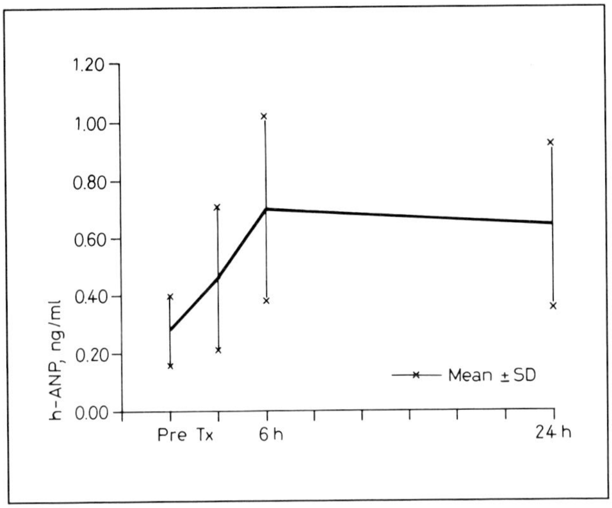

Fig. 1. Pediatric renal transplantation. Plasma concentrations of h-ANP (mean ± SD) before renal transplantation, intraoperatively when renal perfusion was started, and 6 and 24 h after transplantation.

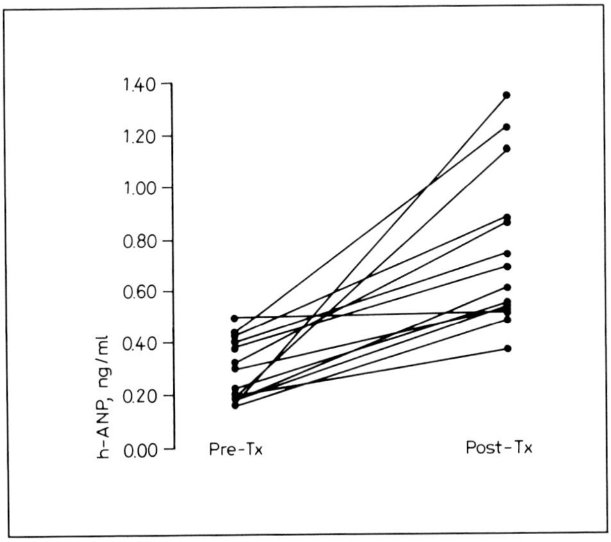

Fig. 2. Pediatric renal transplantation. The individual plasma concentrations of h-ANP (ng/ml) of 14 children before and after renal transplantation.

References

1 Offner, G.; Hoyer, P.F.; Brodehl, J.; Pichlmayr, R.: Cyclosporin A in pediatric kidney transplantation. Pediat. Nephrol. *1:* 125–130 (1987).

2 De Bold, A.J.: Atrial natriuretic factor. A hormone produced by the heart. Science *230:* 767–770 (1985).

3 Cho, S.I.; Bradley, J.W.; Monaco, A.P.; Tilney, N.L.: Comparison of kidney transplant survival between patients treated with cyclosporine and those treated with azathioprine and antithymocyte globulin. Am. J. Surg. *147:* 518–523 (1984).

4 Jueppner, H.; Brabant, G.; Kapteina, U.; Kirscner, M.; Klein, H.; Hesch, R.D.: Direct radioimmunoassay for human natriuretic peptide (hANP) and its clinical evaluation. Biochem. biophys. Res. Commun. *139:* 1215–1223 (1986).

5 Rascher, W.; Tulassay, T.; Lang, R.E.: Atrial natriuretic peptide in plasma of volume-overloaded children with chronic renal failure. Lancet *ii:* 303–305 (1985).

Gisela Offner, MD, Kinderklinik der Medizinischen Hochschule Hannover, Konstanty-Gutschow-Strasse 8, D–3000 Hannover (FRG)

Contr. Nephrol., vol. 67, pp. 53–57 (Karger, Basel 1988)

Atrial Natriuretic Peptide and Other Vasoactive Hormones during Treatment of Severe Diabetic Ketoacidosis in Children[1]

T. Tulassay[a, 2], *W. Rascher*[b], *A. Körner*[a], *M. Miltényi*[a]

[a] Department of Pediatrics, Semmelweis University Medical School, Budapest, Hungary; [b] Department of Pediatrics, University of Heidelberg, FRG

Diabetic ketoacidosis (DKA) is characterized by severe changes in body fluid homeostasis: volume depletion and sodium loss. Hypovolemia and hyponatremia may activate the sympathetic nervous and renin-angiotensin systems and the release of vasopressin. Activation of these hormone systems theoretically should counterbalance the reduced intravascular volume and compensate for sodium loss, thus preventing circulatory insufficiency [1, 2].

Atrial natriuretic peptide (ANP) plays an important role in the regulation of plasma volume and sodium homeostasis [3]. Extracellular volume expansion is associated with an increased ANP secretion [4]. No data are available, however, whether hypovolemia is associated with reduced plasma ANP levels. Children with DKA have severe volume depletion and could be a good clinical model to study the regulation of vasoactive hormones during hypovolemia and after salt and fluid repletion.

The present study was undertaken to determine the alterations of ANP in hypovolemia associated with DKA as well as the interaction with arginine vasopressin, the renin-aldosterone axis, and catecholamines.

[1] Supported by the Deutsche Forschungsgemeinschaft (RA 326/1-2).

[2] T. Tulassay was a recipient of a fellowship of the Alexander von Humboldt-Stiftung, Bonn, FRG.

Patients and Methods

Nine children (6 females) were investigated during DKA. Their ages ranged between 10 and 17 years (mean: 12.6 years) and mean duration of diabetes was 4.5 years (range 0–12.5 years).

All children were severely acidotic; capillary pH ranged between 7.05 and 7.30 (mean: 7.13) and mean base excess was -21 ± 1.9 mmol/l ($\bar{x} \pm$ SEM). Blood glucose concentration reached 608.4 ± 124.2 mg/dl. Insulin (Actrapid MC, Novo) was administered at a dose of 0.1 U/kg BW/h intravenously. A solution of 0.9% NaCl was infused until the blood glucose dropped to 300 mg/dl. Then the infusion was changed for a solution of 0.45% NaCl and 5% glucose. Volume depletion was corrected during the first 24 h. Blood was drawn at 0, 12, 24 and 72 h of the fluid and electrolyte replacement therapy.

Determinations of creatinine, sodium, potassium, phosphate, calcium, and glucose were carried out by conventional laboratory techniques. Osmolality of plasma and urine was measured by freezing point depression (Knauer osmometer).

Plasma concentrations of ANP, plasma renin activity (PRA), aldosterone, arginine vasopressin (AVP) and cortisol were measured by using specific radioimmunoassays, as described elsewhere [4, 5]. Concentrations of norepinephrine in the plasma were determined by a radioenzymatic method [6]. The data are expressed as mean and standard error of the mean ($\bar{x} \pm$ SEM). The Student's paired t test was used for statistical analysis. The correlation coefficient was calculated according to the method of least squares.

Results

Average fluid replacement was $3{,}958 \pm 477$ ml/1.73 m^2 during the first 12 h and $2{,}641 \pm 354$ during the second 12 h ($p < 0.01$). Urine output during these two periods was $2{,}160 \pm 296$ and $1{,}635 \pm 173$ ml/12 h/ 1.73 m^2, respectively.

Blood pH values almost normalized during the first 12 h of therapy and increased further up to the third day. Base excess reached the normal range only on the third day of therapy. Hemoglobin and hematocrit values decreased significantly, as did blood glucose concentrations and plasma osmolality, while the plasma concentration of sodium increased (table I).

Prior to therapy all hormones were markedly abnormal (table II). The ANP level was far below the normal range ($p < 0.01$), whereas plasma concentrations of AVP, PRA, aldosterone, cortisol and norepinephrine were elevated.

Twelve hours after the start of replacement therapy, only the plasma level of cortisol had normalised; ANP, PRA, aldosterone and norepineph-

Table I. Changes in blood pH and base excess, hemoglobin, hematocrit, blood glucose concentration, plasma osmolatity, and plasma concentration of sodium during the course of diabetic ketoacidosis ($\bar{x}$ ± SEM; the pH values are given as median, range)

	Time			
	0 h	12 h	24 h	72 h
pH	7.13 (7.05–7.30)	7.31 (7.13–7.39)	7.35 (7.26–7.45)	7.40 (7.34–7.47)
Base excess, mmol/l	21 ± 1.9	10.4 ± 2.3	9.1 ± 1.2	4.2 ± 1.4
Hemoglobin, g/dl	16.0 ± 0.7	13.7 ± 0.6	14.0 ± 0.4	14.8 ± 0.5
Hematocrit, %	50 ± 1.1	42 ± 1.4	40 ± 1.4	40 ± 0.6
Blood glucose, mg/dl	608.4 ± 124.2	221.4 ± 23.4	241.2 ± 19.8	241.2 ± 30.6
Plasma osmolality, mosm/kg	362 ± 21	288 ± 20	310 ± 14	304 ± 14
Plasma sodium, mmol/l	130 ± 1.4	136 ± 1.7	136 ± 1.7	136 ± 0.5

Table II. Changes in plasma concentrations of ANP, AVP, PRA, aldosterone, cortisol and norepinephrine during the treatment of diabetic ketoacidosis ($\bar{x}$ ± SEM)

	Time			
	0 h	12 h	24 h	72 h
ANP, fmol/ml	5.3 ± 1.2	18.1 ± 3.0[b]	27.6 ± 2.0[d]	26.0 ± 2.2[d]
AVP, pg/ml	158.8 ± 43.5	61.3 ± 15.8[a]	22.2 ± 4.1[b]	20.4 ± 6.8[b]
PRA, ng Ang I/ml/h	58.9 ± 19.2	7.9 ± 2.7[a]	3.28 ± 0.95[a]	1.38 ± 0.18[a]
Aldosterone, ng/dl	114.3 ± 11.3	68.9 ± 10.5[a]	14.0 ± 4.7[d]	6.3 ± 3.1[d]
Cortisol, μg/dl	32.7 ± 2.1	19.8 ± 3.0[c]	16.1 ± 2.3[d]	12.2 ± 1.6[d]
Norepinephrine, pg/ml	430 ± 67	288 ± 31[a]	217 ± 12[b]	189 ± 5.7[c]

[a] $p < 0.05$; [b] $p < 0.02$; [c] $p < 0.01$; [d] $p < 0.001$; compared to baseline values (time '0').

rine concentrations reached the physiologic range after the 24 h, whereas AVP remained elevated throughout the entire study period.

Significant positive correlations were found in DKA between PRA and concentration of aldosterone in the plasma ($n = 9$, $r = 0.75$, $y = 86 + 1.3x$, $p < 0.01$) and between ANP and AVP ($n = 9$, $r = 0.92$, $y = 1.7 + \frac{299}{x}$, $p < 0.001$). Significant indirect correlations were found between changes in plasma concentrations of ANP on the one hand and AVP (r =

0.75, $y = 0.82 + 0.18x$, $p < 0.01$), noradrenaline ($r = 0.64$, $y = 76 + 0.05x$, $p < 0.01$), and PRA ($r = 0.65$, $y = 6.2 + 0.13x$, $p < 0.05$) on the other, during the first 12 h of therapy.

Discussion

In children with diabetic ketoacidosis and severe volume depletion we observed markedly raised activity of several vasoconstrictor hormones, whereas ANP, a vasodilator hormone, was reduced in the plasma. The mechanism triggering renin release appears to be severe volume depletion and sodium loss.

Aldosterone secretion is stimulated in diabetic coma by the renin-angiotensin axis, and also by changes in sodium and potassium homeostasis. The rise of osmotic pressure in the extracellular fluid and volume depletion are the two major stimuli for the release of arginine vasopressin into the circulation [7, 8]. Even after cessation of hypovolemia, the plasma concentration of AVP remained above normal in our patients. The causes of this event are not known.

ANP regulates plasma volume and sodium homeostasis [3, 4]. Our finding of extremely low plasma ANP levels in children with diabetic coma associated with severe volume depletion suggest a negative feedback regulation of that hormone. Thus, hypovolemia and/or decreased atrial pressure inhibit ANP release. Suppression of ANP helps to maintain intravascular volume and prevent further sodium loss.

An inhibitory effect of ANP on AVP, PRA and aldosterone secretion has been demonstrated [9–11], and ANP also inhibits the vasoconstrictor action of norepinephrine [12]. The correlation between changes in ANP values on the one hand and changes in plasma concentrations of AVP, PRA, and norepinephrine on the other, seems to underline the interaction of these hormone systems in the regulation of plasma volume and sodium concentration during diabetic ketoacidosis.

Our data demonstrate that in diabetic ketoacidosis of children the plasma volume and sodium-regulating hormonal systems are markedly altered by volume depletion. Plasma concentrations of vasoconstrictor and sodium- and water-retaining hormones like AVP, renin, aldosterone and norepinephrine are increased, whereas that of ANP with potent vasodilator and natriuretic properties is suppressed. Following volume repletion, most of the alterations are reversible.

References

1 Zerbe, R.L.; Vinicor, F.; Robertson, G.L.: Plasma vasopressin in uncontrolled diabetes mellitus. Diabetes *28:* 503 (1979).
2 Walsh, C.H.; Baylis, P.H.; Malins, J.M.: Plasma arginine vasopressin in diabetic ketoacidosis. Diabetologia *160:* 93 (1979).
3 Ballermann, B.J.; Brenner, B.M.: Biologically active atrial peptides. J. clin. Invest. *76:* 2041 (1985).
4 Rascher, W.; Tulassay, T.; Lang, R.E.: Atrial natriuretic peptide in plasma of volume-overloaded children with chronic renal failure. Lancet *ii:* 303 (1985).
5 Vecsei, P.: Glucocorticoids: cortisol, corticosterone, and compound S; in Jaffe, Behrman, Methods of hormone radioimmunoassays, pp. 393–415 (Academic Press, New York 1974).
6 daPrada, M.; Zürcher, G.: Simultaneous radioenzymatic determination of plasma and tissue adrenaline, noradrenaline and dopamine within the femtomol range. Life Sci. *19:* 1161 (1976).
7 Schrier, R.W.; Berl, T.; Anderson, R.J.: Osmotic and nonosmotic control of vasopressin release. Am. J. Physiol. F *236:* F321 (1979).
8 Rascher, W.; Rauh, W.; Brandeis, W.E.; Huber, K.-H.; Schärer, K.: Determinants of plasma arginine-vasopressin in children. Acta paediat. scand. *75:* 111 (1986).
9 Samson, W.K.: Atrial natriuretic factor inhibits dehydration and hemorrhage-induced vasopressin release. Neuroendocrinology *40:* 277 (1985).
10 Atarashi, K.; Mulrow, P.J.; Franco-Saenz, R.: Effects of atrial peptides on aldosterone production. J. clin. Invest. *76:* 1807 (1985).
11 Burnett, J.C.; Granger, J.P.; Opgenorth, T.J.: Effects of synthetic atrial natriuretic factor on renal function and renin release. Am. J. Physiol. F *247:* F863 (1984).
12 Oshima, T.; Currie, M.G.; Geller, D.M.; Needleman, P.: An atrial peptide is a potent renal vasodilator substance. Circulation Res. *54:* 612 (1984).

T. Tulassay, MD, First Department of Paediatrics, Bókay ut 53,
H–1083 Budapest (Hungary)

Contr. Nephrol., vol. 67, pp. 58–66 (Karger, Basel 1988)

Tubular Dysfunction in Metabolic Acidosis

First Step to Acute Renal Failure[1]

M. Miltényi, T. Tulassay, A. Körner, A. Szabó, M. Dobos

First Department of Paediatrics, Semmelweis University Medical School, Budapest, Hungary

In earlier studies we found reversible tubular proteinuria in diabetic children with hyperglycemic ketoacidosis [1]. This proteinuria is the sign of reversible tubular impairment [2]. Diabetic ketoacidosis is a complex metabolic derangement, characterized mainly by hyperglycemia and metabolic acidosis. Both these conditions may also occur separately or together in pathological conditions other than diabetes. Patients with periodic acetonemic vomiting or with respiratory circulatory failure are normo- or even hypoglycemic, while exhibiting mild-to-pronounced metabolic acidosis. Well-controlled diabetics, as well as poorly controlled diabetics without ketonuria, are characterized by near-normal acid-base balance, even when hyperglycemia appears to be pronounced in the latter group. The aim of the present study was to determine the relationship between the type and degree of proteinuria and that of glycemia and acid-base state in the above-mentioned pathological conditions. Metabolic acidosis due to hypoxia and hypoperfusion is caused by anoxia and is accompanied by exsiccosis in most cases. In the second part of our study we investigated the role of metabolic acidosis due to ammonium-chloride loading. This type of metabolic acidosis is not mixed with hypoxia or hypoperfusion. Therefore, in this group of patients we could investigate the consequences of metabolic acidosis alone.

[1] This research was partly supported by the Alexander von Humboldt-Stiftung, Bonn, FRG.

Patients and Methods

Patients

Seven groups of patients have been investigated: (1) 23 healthy controls; (2) 61 well-controlled diabetics with mild hyperglycemia; (3) 7 poorly controlled diabetics with marked hyperglycemia but without ketoacidosis and ketonuria; (4) 18 patients with acetonemic vomiting; (5) 61 diabetic patients in hyperglycemic ketoacidosis; (6) 27 children suffering from respiratory-circulatory failure with hypovolemic and toxic shock due to enteritis or severe toxic pneumonia. On the basis of their serum creatinine level and urine osmolality, these patients have been divided into two subgroups: (a) 14 children without acute renal failure (ARF), and (b) 13 children suffering from prerenal ARF. Diagnostic criteria were [3]: $S_{Cr} > 150$ μmol/l, $U_{Osm} > 450$ mosm/kg H_2O, no hematuria. (7) Eight children aged 8–13 years with normal renal function, and diagnosis: enuresis nocturna. They were loaded by peroral ammonium chloride to induce metabolic acidosis.

The clinical data are summarized in table I.

Methods

A 24-hour urine specimen was collected in most cases, but in patients with diabetic ketoacidosis (DKA), acetonemic vomiting and respiratory-circulatory failure, the collection period was 4 h. Urinary protein concentration was measured using the method of Yatzidis [4]. The urinary protein pattern (UPP) was determined by sodium dodecyl sulfate polyacrylamide gel electrophoresis (SDS-PAGE) as described previously [5]. The

Table I. Clinical data of the patients

Group	n	Age years	Duration of diabetes years	Blood glucose mmol/l	pH	Base excess mmol/l
Controls	23	10.4 ± 4.4	–	–	7.40 ± 0.02	-1.26 ± 1.4
Well-controlled diabetics	61	12.9 ± 3.6	5.9 ± 3.6	10.67 ± 3.51	7.41 ± 0.04	-2.52 ± 1.45
Hyperglycemic diabetics without ketoacidosis	7	10.8 ± 2.6	recent onset	25.0 ± 4.93	7.39 ± 0.02	-5.1 ± 1.1
Acetonemic vomiting	18	6.63 ± 2.3	–	5.54 ± 1.22	7.34 ± 0.07	-8.5 ± 5.1
Diabetics in hyperglycemic ketoacidosis	61	11.3 ± 3.5	4.1 ± 2.8	29.8 ± 12.5	7.10 ± 0.11	-23.2 ± 4.0
Circulatory-respiratory failure						
(a) Without prerenal ARF	14	1.8 ± 1.5	–	–	7.05 ± 0.05	-18.5 ± 4.6
(b) With prerenal ARF	13	3.2 ± 1.5	–	–	6.95 ± 0.11	-23.5 ± 4.6
Enuresis nocturna, NH_4Cl load	8	8– 13	–	–	see results	

various proteins separated on the gel were grouped as follows: (1) high molecular weight (HMW) globulins with molecular weights of 1,000,000–100,000; (2) albumin; (3) low molecular weight (LMW) microproteins with molecular weights of 50,000–10,000 [6].

Measurement of blood glucose, acid-base balance and urinary ketone bodies was performed by routine laboratory methods. Data of urinary total protein (UTP) excretion, albumin and LMW excretion are expressed as µg/min/1.73 m^2 BSA. In the 7th group of patients a routine NH_4Cl loading test was performed. 0.2 g/kg capsulated NH_4Cl was applied. Investigations were carried out: (1) 2 h before loading; (2) 2 h after; (3) 4 h after; (4) 6 h after, and (5) 26 h after loading.

The following parameters were studied: (1) blood pH, base excess, P_{CO_2}, urinary pH; (2) urinary protein excretion: urinary total protein, albumin, LMW proteins, β_2-microglobulin (β_2M); (3) renal handling of electrolytes: FE_{Na}, FE_K, FE_{Ca}, F_P.

Statistical Analysis

The data of urinary protein excretion are consistent with log-normal distribution [7, 8]. The mean, SEM and linear regression analyses were therefore calculated on logged data. Power calculations were performed with assumptions of unpaired t tests with a significance level of 0.01.

Results

In well-controlled diabetics urinary protein excretion did not differ from that of healthy controls. In poorly controlled (hyperglycemic) diabetics without ketoacidosis and ketonuria, proteinuria was almost entirely normal compared to healthy controls. However, in children with acetonemic vomiting who suffered from metabolic acidosis, total protein excretion was significantly increased due to increased LMW excretion. Diabetics in hyperglycemic ketoacidosis exhibited significantly elevated UTP. According to the percentage excretion of albumin and LMW proteins, proteinuria in this group appeared to be of tubular origin.

In all the children with respiratory-circulatory failure, UTP and LMW protein excretion were significantly increased. UTP levels in those children who suffered from prerenal ARF exceeded those of all other groups included in the study. The results are summarized in table II.

Correlation analysis yielded the following relationship between the urinary excretion of different proteins and the base excess (BE): albuminuria as well as urinary excretion of LMW proteins were both inversely related to BE. Figure 1 shows the correlation between BE and albuminuria. The correlation between BE and LMW proteinuria is depicted in figure 2. On the basis of the correlation coefficient, the relationship of LMW proteinuria to BE was even more pronounced than that of albuminuria.

Table II. UTP output, albumin and LMW excretion (in μg/min/1.73 m^2 BSA) in different patient groups (means ± SEM)

Groups	n	UTP	Albumin	LMW
Controls	23	129[a] 80.9–206.1	69.2[a] 33.2–144.2	9.2[a] 2.5–33.7
Well-controlled diabetics	61	138.4 76.2–251.2	72.6 33.3–158.1	14.3 6.4–32.0
Hyperglycemic diabetics without ketoacidosis	7	165.2 87.7–311.2	72.6 25.6–206.1	27.7 10.2–75.2
Acetonemic vomiting	18	467.7[b] 241.5–905.7	175.8[b] 82.4–375.0	171.8[b] 91.6–322.1
Diabetics in hyperglycemic ketoacidosis	61	1,148.2[b] 582.1–2,264.6	257.0[b] 100.0–660.7	601.2[b] 304.1–1,188.5
Circulatory-respiratory failure				
Without prerenal ARF	14	641.2[b] 430.5–955.0	244.9 152.8–392.6	288.4[b] 181.1–459.2
With prerenal ARF	13	1,399.6[b, c] 682.3–2,870.8	542.0[b, c] 252.9–1,161.4	671.4[b, c] 312.6–1,442.1

[a] Means ± SEM (log-normal).
[b] Significantly different from control, $p < 0.01$.
[c] Significantly different from patients without ARF, $p < 0.01$.

The NH_4Cl load caused a marked metabolic acidosis. The pH dropped from 7.40 ± 0.03 to 7.15 ± 0.08 at the 2nd and 4th h of loading and returned to normal at the 26th h. During the same time the initial base excess decreased from −0.1 to −17.3 ± 1.7 mmol/l ($p < 0.01$). At that time P_{CO_2} went down from 39 ± 5 to 27 ± 8 mm Hg and the urinary pH from 5.8 to 5.0. At the time of metabolic acidosis there was no significant change in UTP excretion or in albumin excretion. The LMW protein excretion was elevated 6 times ($p < 0.02$) and the β_2M 10 times ($p < 0.01$).

There was also a significant alteration in renal salt handling at the time of metabolic acidosis. FE_{Na} rose from 1.5 ± 0.3 to 3.8 ± 0.4% ($p < 0.05$). FE_K elevated from 1.7 ± 0.3 to 4.0 ± 0.5% ($p < 0.05$), and FE_{Ca} from

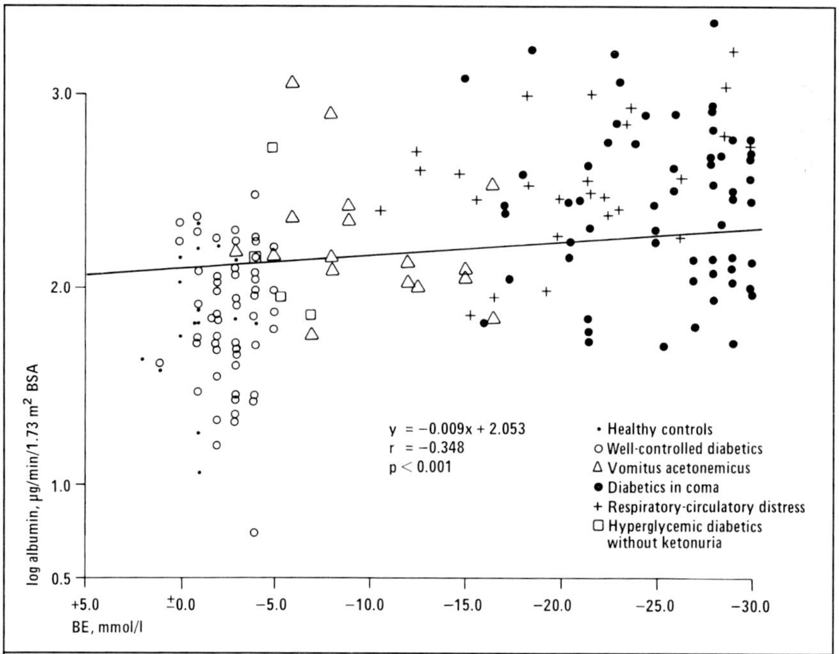
3.0
2.0
1.0
0.5
log albumin, µg/min/1.73 m² BSA
+5.0
±0.0
−5.0
−10.0
−15.0
−20.0
−25.0
−30.0
BE, mmol/l
y = −0.009x + 2.053
r = −0.348
p < 0.001
• Healthy controls
○ Well-controlled diabetics
△ Vomitus acetonemicus
● Diabetics in coma
+ Respiratory-circulatory distress
□ Hyperglycemic diabetics without ketonuria

1

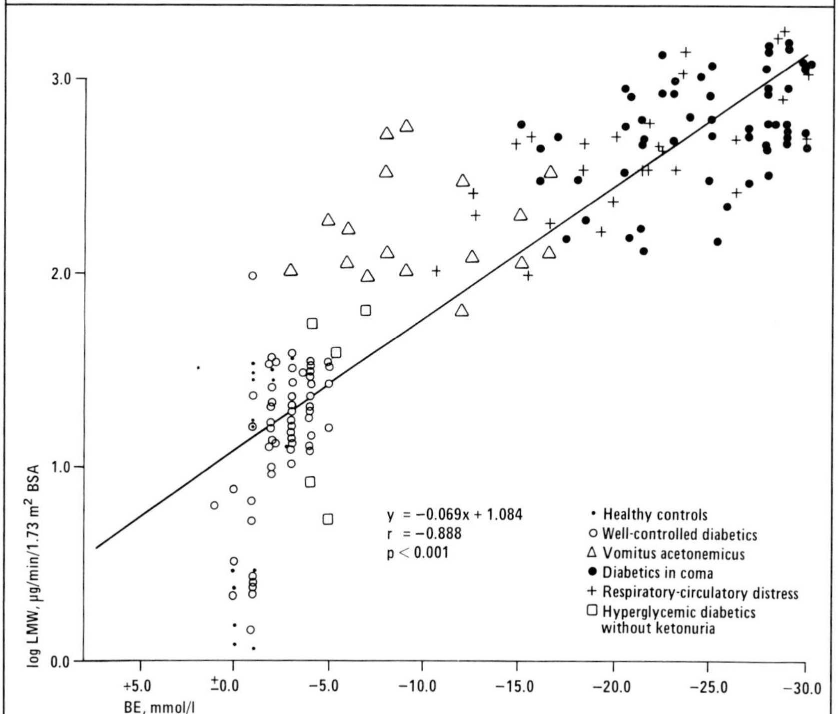
3.0
2.0
1.0
0.0
log LMW, µg/min/1.73 m² BSA
+5.0
±0.0
−5.0
−10.0
−15.0
−20.0
−25.0
−30.0
BE, mmol/l
y = −0.069x + 1.084
r = −0.888
p < 0.001
• Healthy controls
○ Well-controlled diabetics
△ Vomitus acetonemicus
● Diabetics in coma
+ Respiratory-circulatory distress
□ Hyperglycemic diabetics without ketonuria

2

2.9 ± 0.4 to $5.7 \pm 6.0\%$ ($p < 0.05$). The elevation of FE_P was of a smaller degree and not significant.

Therefore, metabolic acidosis without any hypoxia or hypoperfusion caused a transient tubular dysfunction in all the investigated factors.

Discussion

Metabolic acidosis may be the result of circulatory-respiratory failure caused by hypoperfusion and hypoxia, but it can occur in primary metabolic disturbance without hypoxia, as well. Metabolic disturbance affecting the epithelial cells in the proximal tubuli results in decreased reabsorption and/or catabolism of proteins, leading to tubular proteinuria [9, 10].

Previously, we reported on tubular proteinuria in children suffering from hyperglycemic ketoacidosis in type 1 diabetes mellitus [1]. This proteinuria disappears within 7–10 days after successful control of metabolism. Recovery of the impaired tubular function as characterized by FE_{Na} and FE_{Ca}, the tubular phosphate reabsorption rate, and by C_{H_2O} seems to be a slower process [2]. The impaired proximal tubular function could be a consequence of the insufficient glucose utilization [11]. Glucose [12] and lactate [13] are important fuels for respiration in the medulla.

The first aim of our recent study was to determine the distinct effect of the two major components of diabetic ketoacidosis, i.e. hyperglycemia and ketoacidosis, on the type and degree of protein excretion. Hyperglycemia without ketoacidosis does not alter the physiological proteinuria. In these cases the tubular function seems to be maintained by the energy of fatty acids, endogenous lipids and ketone bodies [11]. Even a glucose load exceeding the T_mG of the proximal tubuli does not alter the reabsorption/catabolism of microproteins in the proximal tubuli. However, in patients with periodic acetonemic vomiting, despite normoglycemia, the urinary excretion of tubular microproteins increased significantly. In both pathological states, as judged from the metabolic disturbances, tubular proteinuria is related to metabolic acidosis, regardless of the blood glucose level.

Fig. 1. Correlation between BE and albuminuria.
Fig. 2. Correlation between BE and urinary excretion of LMW protein.

In the second part of our study the goal was to investigate tubular function in metabolic acidosis caused by NH_4Cl load. The induced metabolic acidosis was not accompanied by hypoxia and hypoperfusion. In spite of this, we registered complete tubular dysfunction at the time of metabolic acidosis.

According to our present knowledge, hypoxia-hypoperfusion of the kidney results in profound functional and morphological changes primarily in the proximal tubuli. Whatever the initiating cause of that ARF, the injury at first affects the tubular transport activity [14–22]. Reduction in the O_2 uptake of the kidney decreases both synthetic functions of the cells in the proximal tubuli, as well as Na^+ reabsorption [23]. In the rat, ARF induced by ischemia is associated with marked intracellular acidosis and with a reduction in ATP content of the epithelial cells of the proximal tubuli [24, 25].

Tubular proteinuria can be observed in ARF, as suggested by previous investigations [26]. In the case of reversible ARF, proteinuria can be resolved completely. If the injury is only mild or moderate, recovery of tubular function can be observed after a short period. Severe hypoxic damage leads to tubular necrosis. Recovery, if any, occurs only after a long period. This long recovery phase is characterized by tubular dysfunction, among others by tubular proteinuria.

On the basis of our data, tubular proteinuria occurs in pathological conditions that lead to pronounced metabolic acidosis – regardless of their origin. There is an inverse relationship between the degree of tubular proteinuria and the BE. Thus, metabolic acidosis is accompanied by tubular impairment even in nonanoxic states, such as diabetic ketoacidosis, or NH_4Cl load.

In conclusion, metabolic acidosis seems to play a key role in tubular impairment, and can be regarded as a first step in hypoxia-induced ARF. Tubular proteinuria is a sensitive and early indicator of this process.

References

1 Miltényi, M.; Körner, A.; Dobos, M.: Tichy, M.: Reversible tubular proteinuria associated with hyperglycemic ketoacidosis in type 1 diabetes mellitus. Int. J. pediat. Nephrol. *4:* 247–250 (1983).

2 Miltényi, M.; Körner, A.; Tulassay, T.; Szabó, A.: Tubular dysfunction in type I diabetes mellitus. Archs Dis. Childh. *60:* 929–931 (1985).

3 Anand, S.K.; Northway, J.D.; Crussi, F.G.: Acute renal failure in newborn infants. J. Pediat. *92:* 985–988 (1978).

4 Yatzidis, H.: New colorimetric method for quantitative determination of protein in urine. Clin. Chem. *23:* 811–812 (1977).

5 Miltényi, M.; Pohlandt, F.; Bóka, G.; Kun, E.: Tubular proteinuria after perinatal hypoxia. Acta paediat. scand. *70:* 399–403 (1981).

6 Miltényi, M.: Urinary protein excretion in healthy children. Clin. Nephrol. *12:* 216–221 (1979).

7 Savory, J.; Pu, P.H.; Sunderman, W., Jr.: A biuret method for determination of protein in normal urine. Clin. Chem. *14:* 1160–1171 (1968).

8 Barratt, T.M.; McLaine, P.N.; Soothill, J.F.: Albumin excretion as a measure of glomerular dysfunction in children. Archs Dis. Childh. *45:* 496–501 (1970).

9 Maack, T.; Johnson, V.; Kau, S.T.; Figueiredo, J.; Sigulem, D.: Renal filtration, transport, and metabolism of low-molecular-weight proteins: a review. Kidney int. *16:* 251–270 (1979).

10 Carone, F.A.; Peterson, D.R.; Oparil, S.; Pullman, T.N.: Renal tubular transport, catabolism of proteins and peptides. Kidney int. *16:* 271–278 (1979).

11 Ross, B.D.; Espinal, J.; Silva, P.: Glucose metabolism in renal tubular function. Kidney int. *29:* 54–67 (1986).

12 Eveloff, J.; Haase, W.; Kinne, R.: Separation of renal medullary cells: isolation of cells from the thick ascending limb of Henle's loop. J. Cell Biol. *87:* 672–680 (1980).

13 Bartlett, S.; Lowry, M.; Ross, B.D.: Competition between lactate and glucose in the renal medulla. Proc. 8th Int. Congr. Nephrol, Athens 1981, RM.054, p. 66.

14 Thurau, K.; Boylan, J.W.: Acute renal success. The unexpected logic of oliguria in acute renal failure. Am. J. Med. *61:* 308–315 (1976).

15 Thurau, K.; Beck, J.; Muson, R.; Dörge, A.: Intracellular electrolytes of proximal and distal tubular cells following renal ischemia; in Eliahou, Acute renal failure, pp. 10–14 (Libbey, London 1982).

16 Myers, B.D.; Miller, D.C.; Mehigan, J.T.; Olcott, C.; Golbetz, H.; Robertson, C.R.; Derby, G.; Spencer, R.; Friedman, S.: Nature of renal injury following total renal ischemia in man. J. clin. Invest. *73:* 329–341 (1984).

17 Parekh, N.; Esslinger, H.; Steinhausen, M.: Glomerular filtration and tubular reabsorption in post-ischemic acute renal failure. Kidney int. *25:* 33–41 (1984).

18 Brezis, M.; Rosen, S.; Silva, P.; Epstein, F.H.: Renal ischemia: a new perspective. Kidney int. *26:* 375–378 (1984).

19 Brezis, M.; Rosen, S.; Silva, P.; Epstein, F.H.: Selective vulnerability of the medullary thick ascending limb to anoxia in the isolated perfused rat kidney. J. clin. Invest. *73:* 182–190 (1984).

20 Shanley, P.F.; Brezis, M.; Spokes, K.; Silva, P.; Epstein, F.H.; Rosen, S.: Hypoxic injury in the proximal tubule of the isolated perfused rat kidney. Kidney int. *29:* 1021–1032 (1986).

21 Shanley, P.F.; Brezis, M.; Spokes, K.; Silva, P.; Epstein, F.H.; Rosen, S.: Transport-dependent cell injury in the S_3 segment of the proximal tubule. Kidney int. *29:* 1033–1037 (1986).

22 Myers, B.D.; Moran, S.M.: Hemodynamically mediated acute renal failure. New Engl. J. Med. *314:* 97–105 (1986).

23 Cohen, J.J.: Relations between energy requirements for Na^+ reabsorption and other renal functions. Kidney int. *29:* 32–40 (1986).
24 Chan, L.; Ledingham, J.G.G.; Clarke, J.; Ross, B.D.: The importance of pH in acute renal failure; in Eliahou, Acute renal failure, pp. 58–61 (Libbey, London 1982).
25 Ratcliffe, P.J.; Moonen, C.T.W.; Holloway, P.A.H.; Ledingham, J.G.G.; Radda, G.K.: Acute renal failure in hemorrhagic hypotension. Cellular energetics and renal function. Kidney int. *30:* 355–360 (1986).
26 Revillard, J.P.; Man, E.L.Y.; François, R.; Traeger, J.: Renal diseases associated with tubular proteinuria; in Manuel, Revillard, Beutel, Proteins in normal and pathological urine, p. 213 (University Park Press, Baltimore 1970).

M. Miltényi, MD, First Department of Paediatrics,
Semmelweis University Medical School, Bókay J.u. 53, H–1083 Budapest (Hungary)

Contr. Nephrol., vol. 67, pp. 67–70 (Karger, Basel 1988)

Free Cytosolic Calcium in Children with Idiopathic Hypercalciuria

L. Ghio[a], *M.L. Bianchi*[b], *V. Cecchetti*[a], *L. Soldati*[b], *S. Ortolani*[b], *M. Bacchini*[a], *A. Edefonti*[a], *F. Sereni*[a]

[a]Clinica Pediatrica G. e D. De Marchi-Università Milano;
[b]Istituto Scienze Mediche-Università Milano, Italia

The pathogenesis of idiopathic hypercalciuria (IH) is a puzzling problem. Increased intestinal calcium absorption, calcium renal leakage [1], disordered regulation of vitamin D_3 synthesis [2, 3] and altered bone resorption [4] have all been considered as possible causes. However, it is still not possible to define the exact pathogenesis of IH, despite the large amount of experimental and clinical data which have accumulated.

Some authors have used a new biochemical-cellular approach to study the pathogenesis of IH, and an abnormal elevated erythrocyte (Ca-Mg)ATPase activity has recently been found in adult hypercalciuric patients [5].

Based on these data, we studied free cytosolic calcium, one of the most important pools of intracellular calcium, in the lymphocytes of hypercalciuric pediatric patients. The ability to measure intracellular calcium in these cells may be an important prerequisite for understanding calcium homeostasis in renal tubular cells and enterocytes.

Patients and Methods

The study was carried out on 13 children (8 female, 5 male; aged 4–11 years, mean 6.6 ± 2.2) in whom a diagnosis of IH had been made in our department during the preceding 2 years. The diagnosis was established on the basis of urinary calcium excretion rate > 4 mg/kg/day [6] (6.1 ± 1.3 mg/kg/day in our population) and/or calcium/creatinine ratio (mg/mg) > 0.21 [7] (0.27 ± 0.07 in our population), with normal serum calcium. Hypercalciuria was confirmed in at least three 24-hour urine samples, collected while the children were on a free diet. None of the patients took medications during our study. To exclude other causes of hypercalciuria, we measured the glomerular filtration rate, serum phosphorus, blood acid-base parameters, alkaline phosphatase, and renal tubular phosphate threshold and performed radiological investigations. Twelve patients were referred to us with microscopic or gross hematuria. Two of the patients with gross hematuria passed renal stones. One child was asymptomatic and elevated urinary calcium excretion was found when she was investigated because her sister had hypercalciuria.

Free cytosolic calcium was determined with a new fluorescent calcium indicator (Quin 2) [8] which enters intact cells in the form of a hydrophobic membrane-permeable ester and is then trapped inside after hydrolysis. Lymphocytes, separated by adequate centrifugation, were used for the test. Quin 2 was added to the lymphocyte suspension and then incubated for 60–90 min to allow for hydrolysis of the ester. Fluorescence was measured with a Kontron Instruments spectrofluorometer (SFM25). Twelve healthy children, aged 2–12 years (mean 6.2 ± 1.1), with no history of renal disease, were used as a control group. The following calciotropic hormones were determined: plasma immunoreactive parathyroid hormone mid-molecule (Immunonuclear Co., RIA kit) and amino-terminal (Nichols, RIA kit); plasma immunoreactive calcitonin (Nichols, RIA kit); plasma 25(OH)-vitamin D (IRE, protein-binding assay kit); plasma 1,25$(OH)_2$-vitamin D (Immunonuclear Co, RIA kit). Seventy-two healthy children, aged 4–12 years (mean 8.2 ± 1) were used as the control group.

Results are expressed as means ± 1 SD. Statistical analysis was performed by Student's unpaired t test.

Results

The data for free cytosolic calcium are shown in figure 1. The free cytosolic calcium concentration was significantly lower in the hypercalciuric children (84.8 ± 13.2 nmol/l) in comparison with the controls (105.5 ± 15.2 nmol/l) ($p < 0.005$). No significant differences were found between the hypercalciuric patients and the controls for the calciotropic hormones.

Comment

Although no alterations of conventional parameters of calcium metabolism were found, the finding of a lower cytosolic calcium concentration in lymphocytes of children with IH is in favour of the existence in IH of a derangement of cellular calcium metabolism. In fact, recent experimental observations have shown that the cytosolic calcium concentration in lymphocytes could be considered as a marker of kidney total calcium [9].

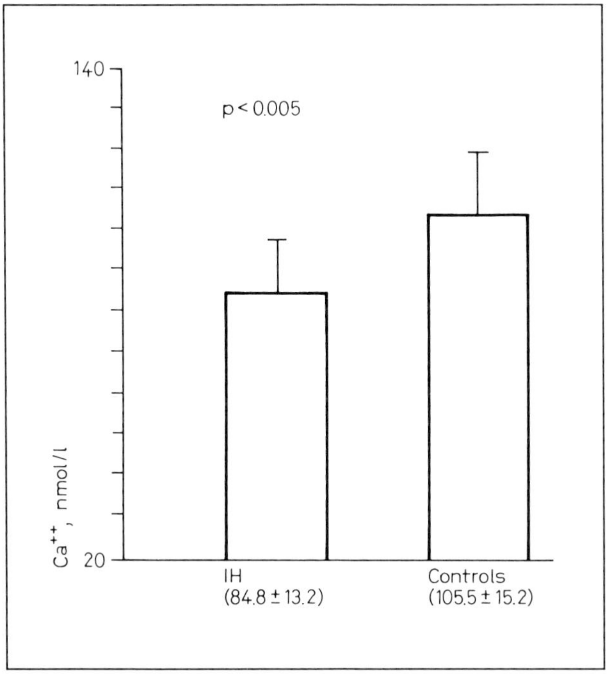

Fig. 1. Free cytosolic calcium concentration in patients and controls.

However, many other aspects should be studied. In particular, the study of calcium pump activity, in order to establish a relation between cytosolic calcium concentration and calcium ion transport, could help clarify the pathogenesis of IH in children.

References

1 Pack, C.Y.C.; Ohata, M.; Lawrence, E.C.; Snyder, W.: The hypercalciurias: causes, parathyroid function and diagnostic criteria. J. clin. Invest. *57:* 387–400 (1974).

2 Broadus, A.E.; Insogna, K.L.; Lang, R.; Ellison, A.F.; Dryer, B.E.: Evidence for disordered control of 1,25-dihydroxyvitamin D production in absorptive hypercalciuria. New Engl. J. Med. *311:* 73–80 (1984).

3 Stapleton, F.B.; Langman, C.B.; Bittle, J.; Miller, L.A.: Increased serum concentration of 1,25$(OH)_2$ D in children with fasting hypercalciuria. J. Pediat. *110:* 234–237 (1987).

4 Sutton, R.A.L.; Walker, V.R.: Bone resorption and hypercalciuria in calcium stone formers. Metabolism *35:* 485–488 (1986).

5 Vezzoli, G.; Elli, A.; Palazzi, P.; Menghetti, D.; Soldati, L.; Ortolani, S.; Quarto di Palo, F.; Surian, M.; Bianchi, G.: (Ca-Mg)ATPase activity in erithrocite membrane ghosts in idiopathic hypercalciuria. Mineral Metab. Res. Italy *6:* 25–27 (1985).

6 Ghazali, S.; Barrat, T.M.: Urinary excretion of calcium and magnesium in children. Archs Dis. Childh. *49:* 97–101 (1974).

7 Stapleton, F.B.; Noe, N.; Jerkins, G.; Roy, S., III: Urinary excretion of calcium following an oral calcium loading test in healthy children. Pediatrics, Springfield *69:* 594–597 (1982).

8 Tsien, R.Y.; Pozzan, T.; Rink, T.J.: Calcium homeostasis in intact lymphocytes: Cytoplasmic free calcium monitored with a new, intracellular trapped fluorescent indicator. J. Cell Biol. *94:* 325–334 (1982).

9 Tsao, P.; Radde, I.C.; Klip, A.: Furosemide increased both intracellular free calcium in lymphocytes as well as total calcium in renal tissue. Abstr. IXth Int. Conf. on Calcium Regulating Hormones and Bone Metabolism, Nice 1986, p. 288.

L. Ghio, MD, Nefrologia e Dialisi, Clinica Pediatrica II, Università Milano,
Via Commenda 9, I–20122 Milano (Italy)

Contr. Nephrol., vol. 67, pp. 71–74 (Karger, Basel 1988)

Renal Dopamine-1 Defect in Spontaneous Hypertension[1]

Robin A. Felder[a], *Mouin G. Seikaly*[b], *Gilbert M. Eisner*[b], *Pedro A. Jose*[b]

[a] University of Virginia Medical Center, Charlottesville, Va.;
[b] Georgetown University Medical Center, Washington, D.C., USA

The sympathetic nervous system, via renal efferent nerves, contributes to the development of hypertension in the spontaneously hypertensive rat (SHR) in part by increasing sodium retention [1]. Renal alpha-adrenoceptors increase [1], while dopamine (DA) receptors decrease sodium transport [2]. However, in the SHR the renal concentrations of norepinephrine are similar to normotensive controls while renal concentrations of DA are increased [3]. The DA-1 receptor has been reported in the proximal convoluted tubule [4] where abnormalities in the renal handling of sodium have been localized [5]. Acute unilateral renal denervation (DNX) induces a natriuresis to a greater extent in SHR than normotensive controls [1]. In normotensive rats, the natriuresis is in part due to a decrease in renal sympathetic nerve input and norepinephrine excretion and an increase in DA excretion [6]. It is therefore possible that the abnormal sodium handling in SHR may be related to increases in renal nerve activity, decreased dopaminergic activity, or both.

In order to evaluate the interaction of these systems in sodium excretion in SHR, we studied the effects of the intrarenal infusion of the benzazepine DA-1 agonist, SKF 38393 [7] in SHR and normotensive Wistar-Kyoto rats (WKY) with renal nerves intact or after DNX. In addition, renal DA-1 receptor density, affinity, and distribution were studied by radioligand binding.

[1] This study was supported in part by National Institutes of Health Grants DK 39308, HL 33498, and USPHS training grant HL 07213.

Materials and Methods

Male SHR of the Aoki-Okamoto strain and WKY serving as controls (age 11–16 weeks, Harlan-Sprague Dawley, Indianapolis, Ind.) were anesthetized, prepared for sham or unilateral DNX and renal clearances [2]. A PE tubing was introduced into the left carotid artery and into the left renal artery via the descending aorta. Thereafter, an intrarenal infusion of heparinized saline was started at 0.5 ml/h. After an equilibrium period of 90 min, a 40-min control urine collection was obtained; thereafter the animals underwent sham or acute left chemosurgical DNX [2]. The glomerular filtration rate (GFR) was determined by the clearance of ^{3}H-inulin. After the control periods the animals received an intrarenal infusion of DA-1 agonist SKF 38393 (1.2, 12 and 120 ng/g kidney weight/min). A 40-min urine collection period was obtained 10 min after each change of drug concentration and after stopping the drug infusion, designated as the recovery period. In another group of WKY, the specificity of SKF 38393 to the DA-1 receptor was tested by the infusion of the DA-1 antagonist SCH 23390 (1.2, 12 and 120 ng/g kidney weight/min). DA-1 receptor affinity and density were measured by specific binding of the DA-1 radioligand ^{125}I-SCH 23982 (defined by 5 μM SCH 23390). DA-1 receptors were localized by autoradiography.

Results and Discussion

In WKY with renal nerves intact, infusion of SKF 38393 had no effect on mean arterial pressure (MAP) or GFR but increased urine flow (V) and absolute ($U_{Na}V$) and fractional sodium excretion (Fe_{Na}) in response to the highest two doses of SKF 38393. Recovery to preinfusion levels occurred after stopping the drug. The contralateral kidney was unaffected, indicating absence of significant spillover into the systemic circulation. Intrarenal arterial infusion of vehicle alone did not influence ipsilateral or contralateral renal function. In contrast to the results in WKY, intrarenal administration of SKF 38393 had no effect on MAP, GFR, V, $U_{Na}V$ or Fe_{Na} in SHR with intact renal nerves. SKF 38393 exerted its action via the DA-1 receptor, since SCH 23390 alone had no effect on any of the parameters studied but did block the natriuretic and diuretic effects of SKF 38393. The decreased dopaminergic response noted in SHR was not due to downregulation of DA-1 receptors induced by the high renal concentrations of dopamine in SHR [3]. DA-1 receptor affinity and density in renal cortical homogenates and location by autoradiography in kidney slices were similar in WKY and SHR.

The inability of SKF 38393 to induce natriuresis and diuresis in SHR could have been due to masking of DA-like activity by increased renal nerve activity [1]. Acute DNX increased V and $U_{Na}V$ to a greater extent in

SHR than WKY without affecting MAP or GFR. After acute renal DNX, SKF 38393 increased GFR in WKY but not SHR. This suggests that in normotensive rats tonic renal adrenergic nerve activity may act to dampen the dopaminergic effects on the glomerular dynamics; this effect is not seen in SHR. After acute DNX in WKY, SKF 38393 increased V and $U_{Na}V$ to a greater extent than in innervated WKY kidneys. In SHR, acute DNX unmasked a diuretic and natriuretic effect of SKF 38393. However, the effects were still less than in the denervated WKY when expressed as change from post-DNX values. This suggests again an impaired response of the SHR to DA stimulation. In the rat, renal nerve stimulation leads to antinatriuresis apparently due in part to alpha-adrenergic effects; blockade of alpha-adrenoceptors during renal nerve stimulation leads to a natriuresis, apparently due to dopaminergic activity [8]. The natriuresis and diuresis of acute unilateral DNX in the rat is associated with increased DA excretion [6]; blockade of DA-1 receptors attenuated the natriuresis [2]. The increased renal nerve activity in SHR [1] in association with decreased renal dopaminergic effects [9] may explain the sodium retention in SHR [1].

Conclusion

Occupation of rat renal DA-1 receptors results in a natriuresis in normotensive rats that was enhanced after acute unilateral renal denervation. In addition, DA-1 agonist increased the GFR in normotensive rats only after denervation. In the SHR, DA agonists increased sodium excretion only after denervation and then to a lesser extent than normotensive controls and without affecting the GFR. The decreased natriuretic response to DA agonists is not due to differences in renal DA-1 receptor density or distribution. These studies suggest that the abnormal renal sodium handling in SHR is both due to increased renal nerve ativity and decreased post-DA-1 receptor activity.

References

1 Koepke, J.; Kopp, U.; DiBona, G.: The kidney in the pathogenesis of hypertension: role of the renal nerves; in Kaplan, Brenner, Laragh, Perspectives in hypertension. The kidney in hypertension, pp. 53–65 (Raven Press, New York 1987).

2 Jose, P.; Felder, R.; Holloway, R.; Eisner, G.: Dopamine receptors modulate sodium excretion in denervated kidney. Am. J. Physiol. F *250:* F1033–F1038 (1986).
3 Racz, K.; Kuchel, O.; Buu, N.; Tenneson, S.: Peripheral dopamine synthesis and metabolism in spontaneously hypertensive rats. Circulation Res. *58:* 889–897 (1986).
4 Felder, R.; Blecher, M.; Calcagno, P.; Jose, P.: Dopamine receptors in the proximal tubule of the rabbit. Am. J. Physiol. F *247:* F499–F505 (1984).
5 Parenti, P.; Hanozet, G.; Bianchi, G.: Sodium and glucose transport across renal brush border membranes of Milan hypertensive rats. Hypertension *8:* 932–939 (1986).
6 Morgunov, N.; Baines, A.: Renal nerves and catecholamine excretion. Am. J. Physiol. F *240:* F75–F81 (1981).
7 Molloy, S.; Waddington, J.: Dopaminergic behaviour stereospecifically promoted by the D_1 agonist R-SK&F 38393 and selectively blocked by the D_1 antagonist SCH 23390. Psychopharmacology *82:* 409–410 (1982).
8 Chapman, B.; Munday, K.; Radhi, A.: The effects of renal nerve stimulation in rats treated with a-adrenergic and dopaminergic blockers (Abstract). J. Physiol., Lond. *341:* 65p (1983).
9 Lee, M.: Dopamine and the kidney. Clin. Sci. *62:* 439–448 (1982).

R.A. Felder, PhD, University of Virginia Medical Center,
Department of Pathology, Box 168, Charlottesville, VA 22908 (USA)

Contr. Nephrol., vol. 67, pp. 75–78 (Karger, Basel 1988)

Possible Genetic Role of Red Cell Membrane Electrolyte Transport Abnormalities in Families with Essential Hypertension[1]

J.E. Deal, V. Shah, G. Goodenough, M.J. Dillon

Department of Paediatric Nephrology, Institute of Child Health, London, UK

In recent years there has been a growing realisation that a disturbance of the transmembrane distribution of ions is a feature of hypertension in man and that this disturbance can be studied in cells not involved in pressure regulation as well as in those participating directly [1]. In adults, studies have looked at the distribution of the sodium ion and, in particular, its distribution in relation to the white blood cell [2].

Studies in hypertension have shown that the intracellular sodium concentration is high when compared to non-hypertensive controls, and that there are abnormalities of the active pathways of sodium transport; the sodium pump [3], countertransport [4] and cotransport systems [5].

In paediatrics, because of the large volumes of blood required to study the changes in white blood cells, we have chosen to study the red blood cell and the activity of the sodium-potassium ATPase-dependent pump. By using a miniaturised method we have been able, on 6 ml of blood, to measure the intracellular sodium concentration (IcNa), the sodium efflux rate constant (Na22ERC) and the binding of labelled ouabain to the pump sites on the red cell membrane (BMax). Previously, we have found a significant difference between children with primary hypertension and normal controls or children with secondary hypertension [6]. These changes may have an aetiological role or could be just genetic markers.

In this study, using the same methods, we have looked at the families of children presenting to our unit with hypertension.

[1] Financial support was provided by the Charlotte Parkinson Research Fund, John Herring and Friends (Child Health Research Appeal Trust) and the Kidney Research Aid Fund.

Patients and Methods

Families of 23 children presenting for investigation of hypertension were studied. Blood pressure was measured using a random zero sphygmomanometer and the mean of three measurements was recorded for each individual. A urine specimen was obtained for urinalysis and measurement of sodium and creatinine concentrations; and a heparinised blood sample was taken from each member of the family for sodium transport studies, electrolytes and creatinine estimation.

Intracellular sodium concentration (IcNa) was measured by flame photometry, pump activity was estimated by determining the sodium efflux rate constant using sodium 22 (Na22ERC) and the number of pump sites were measured using the specific inhibitor ouabain (BMax).

Table I. All children: results expressed as mean ± 1 SD

	Positive family history	Negative family history
IcNa, mmol/l cells	8.36±2.22	6.26±1.14 ($p < 0.001$)
Na22ERC	0.4676±0.1121	0.5771±0.115 ($p < 0.005$)
BMax, nmol/l cell	7.65±1.65	9.24±1.85 ($p < 0.001$)

Table II. Parents: results expressed as mean ± 1 SD

	Hypertensive	Normotensive
IcNa, mmol/l cells	8.40±2.74	6.13±1.30 ($p < 0.01$)
Na22ERC	0.4622±0.0981	0.5913±0.1245 ($p < 0.001$)
BMax, nmol/l cells	8.40±1.88	10.54±2.58 ($p < 0.005$)

Table III. Normotensive siblings: results expressed as mean ± 1 SD

	Positive family history	Negative family history
IcNa, mmol/l cells	8.67±2.39	5.73±1.12 ($p < 0.005$)
Na22ERC	0.4408±0.0986	0.5543±0.1052 ($p < 0.05$)
BMax, nmol/l cells	7.69±1.50	9.9±2.31 ($p < 0.05$)

Results

Fourteen families were identified in which one or other parent had been diagnosed as having essential hypertension or had a definite positive family history of hypertension. Nine families did not have a positive hypertensive family history.

Twenty-eight children from families with a positive family history were compared with the 16 children from the negative family history group. Significant differences were found between these two groups in terms of their sodium transport studies, as shown in table I.

Fifteen of 42 parents had essential hypertension or a positive hypertensive family history and differed significantly from the 27 normotensive adults (table II).

When the index hypertensive children were excluded from the analyses similar significant differences were still found between the 14 children with and the 6 children without a hypertensive family history (table III).

Positive and negative family history groups did not differ significantly in terms of age, blood pressure, renal function or urinary sodium excretion.

Conclusion

These results show that a family history of hypertension significantly influences red blood cell sodium handling. Of particular interest are the results from the normotensive siblings investigated. The findings, in these children, may be the earliest abnormality in the development of essential hypertension and may provide a screening test for identifying normotensive children of hypertensive families at risk of developing hypertension.

References

1 Aalkjaer, C.; Parvin, S.D.; Bing, R.F.; Heagerty, A.M.; Bell, P.R.F.; Swales, J.D.: Cell membrane sodium transport. A correlation between human resistance vessels and leucocytes. Lancet *i:* 649–651 (1986).

2 Edmondson, R.P.S.; Thomas, R.D.; Hilton, P.J.; Patrick, J.; Jones, N.F.: Abnormal leucocyte composition and sodium transport in essential hypertension. Lancet *i:* 1003–1005 (1975).

3 Postnov, Y.V.; Orlov, S.N.; Shevchenko, A.; Adler, A.M.: APTase activity in the red blood cell membrane in essential hypertension. Pflügers Arch. *371:* 263–269 (1977).

4 Canessa, N.; Adragna, N.; Solomon, H.S.; Connoly, T.M.; Tusteson, D.C.: Increased sodium-lithium countertransport in red cells of patients with essential hypertension. New Engl. J. Med. *302:* 772–776 (1980).

5 Garay, R.P.; Elghozi, J.L.; Dagher, G.; Meyer, P.: Laboratory distinction between essential and secondary hypertension by measurement of erythrocyte cation fluxes. New Engl. J. Med. *302:* 769–771 (1980).

6 Uchiyama, M.; Daman-Willems, C.E.; Shah, V.; Dillon, M.J.: Sodium transport in erythrocytes: differences between normal children and children with primary and secondary hypertension. Clin. exp. Hypertens. A*8:* 669–671 (1986).

J.E. Deal, MRCP, Department of Paediatric Nephrology, Institute of Child Health, University of London, 30 Guilford Street, London WC1N 1EH (UK)

Contr. Nephrol., vol. 67, pp. 79–85 (Karger, Basel 1988)

Connective Tissue Metabolism in Children with Chronic Renal Failure

L. Klujber[a], *S. Túri*[b], *I. Haszon*[b], *E. Sulyok*[c]

Departments of Pediatrics, University Medical School of [a]Pécs and [b]Szeged; [c]County Children's Hospital, Pécs, Hungary

Chronic renal failure (CRF) is usually associated with impaired endocrine functions, complex metabolic disturbances and growth failure. The involvement of disturbed connective tissue metabolism in these processes is not completely understood. However, there have been reports of changes in collagen metabolism. Collagen has been shown to be degraded to peptide-bound and free hydroxyproline (HYP) which are excreted in the urine. Urinary peptide-HYP excretion, therefore, has been suggested to be a reliable estimate of collagen turnover and growth activity [1, 2]. In CRF, a low rate of collagen turnover with reduced urinary excretion of HYP peptides has been demonstrated [3].

Besides collagen, polyanionic glycosaminoglycans (GAGs), the polysaccharide-containing elements of proteoglycans, are the major components of connective tissue and form its ground substance. Data on the metabolism of GAGs in CRF are only fragmentary, although elevated circulating hyaluronate has recently been reported and it has been considered as a potential marker of altered connective tissue metabolism in uremic adults [4].

The objective of the present study was to determine whether, in children with advanced CRF, the previously described disturbances of collagen metabolism are accompanied by any alterations of urinary excretions of total and individual GAG compounds. In addition, an attempt was made to relate the pattern of urinary GAG excretion to protein nutrition, severity of CRF and growth. Furthermore, the effect of hemodialysis on GAG excretion was also estimated.

Table I. Clinical data and biochemical parameters of children with CRF

Patient	Sex	Age years	Height cm	Weight kg	Diagnosis	Treatment	Total protein, g/l	Albumin g/l	Urea μ*M*/l	Creatinine μ*M*/l
B.Zs.	F	13	140	34.2	FSGS	hemodialysis	60	28	22.9	844
B.M.	F	17	138	31.5	RH	hemodialysis	64	41	31.7	639
N.E.	F	17	157	43.7	RH	hemodialysis	73	41	31.9	893
F.A.	M	7	113	20.8	FSGS	hemodialysis	33	20	18.1	446
H.Z.	M	13	149	47	H	hemodialysis	72	38	29.2	530
K.N.	F	15	146	32	NC	conservative	80	44	29	467
Sz.L.	M	10	126	28	RN	conservative	68	40	17.1	271
W.A.	M	13	131	36	OU	conservative	69	38	18.7	280
T.P.	M	17	185	74	H	conservative	67	42	10.4	202
K.Zs.	M	9	135	36.5	RH	conservative	74	40	17.2	215

FSGS = Focal segmental glomerulosclerosis; RH = renal hypoplasia; H = hydronephrosis; NC = nephrocalcinosis; RN = reflux nephropathy; OU = obstructive uropathy.

Patients and Methods

Ten children with advanced CRF were selected for the study. Clinical data and some biochemical parameters are listed in table I. Five patients had been on regular hemodialysis for more than 6 months and consumed a normal diet during the study. The other patients had been treated conservatively with a diet moderately restricted in protein (0.7 g/kg/day). Any disturbances in fluid balance, electrolytes and acid-base status or hypertension were treated appropriately. Studies were performed once in nondialyzed children and repeatedly in those receiving hemodialysis. Ten age-matched healthy children served as controls.

After an overnight fast blood samples were taken for amino acids, total protein, albumin, urea and creatinine concentration measurements. All samples were centrifuged immediately and frozen at –20 °C until analyzed. Urine was collected fractionally and samples kept at 4 °C without adding any preservative. After completing the collection it was acidified to pH 5 with 1 *N* hydrochloric acid and kept at –20 °C until analysis.

Laboratory Procedures

Plasma urea, creatinine and protein were measured using standard laboratory procedures. Plasma amino acids were analyzed with a Beckmann multichrome autoanalyzer with lithium buffer system. Assays of urinary free and total HYP were performed by the method of Kivirikko et al. [5] on double samples of all urine specimens. Urinary 3-methyl-histidine (3-met-HIS) was measured colorimetrically according to the method of Radha and Bessman [6].

Isolation and fractionation of urinary GAGs were done after cetylpirydinium chloride (CPC) precipitation on a cellulose celite column. The GAG-CPC-celite complex was then fractionated by the method of Tanaka and Gore [7]. Uronic acid was estimated by the carbazole reaction in every effluent tube [8]. In fractions containing KS (no uronic acid present) galactose was determined by the orcinol method [8]. Total GAG excretion was calculated as the sum of the quantity of uronic acid and of galactose-containing fractions given as µ*M* uronic acid/day.

Statistical Analysis

Results are presented as the mean ± the standard error of the mean ($\overline{X}$ ± SEM). Statistical analysis was carried out by calculating the correlation coefficient, linear regression equation, and Student's t test.

Approval of the institutional ethical committee and informed parental consent was obtained for the study.

Results and Comments

Amino Acids

Plasma concentrations of individual amino acids in patients with advanced CRF and in healthy children of comparable age are shown in table II. Since plasma amino acid concentrations in nondialyzed CRF

Table II. Plasma amino acid patterns in healthy children and in children with CRF before and after hemodialysis (μ*M*/l)

	Control	Before dialysis	After dialysis
Taurine	49 ± 32	82 ± 29	29 ± 3
Asparagic acid	39 ± 12	139 ± 19**	65 ± 21
Threonine	139 ± 32	169 ± 38	76 ± 25
Serine	142 ± 29	207 ± 26	95 ± 21
Glutamic acid	596 ± 66	169 ± 25**	110 ± 40
Proline	187 ± 58	39 ± 11**	22 ± 4
Glycine	232 ± 36	296 ± 53	193 ± 56
Alanine	361 ± 88	387 ± 53	201 ± 53
Valine	223 ± 31	192 ± 61	97 ± 28
Isoleucine	67 ± 13	94 ± 18	57 ± 11
Leucine	127 ± 21	137 ± 39	73 ± 18
Thyrosine	67 ± 11	55 ± 15	29 ± 12
Phenylalanine	58 ± 8	77 ± 16	52 ± 8
Ornithine	49 ± 14	166 ± 70**	65 ± 8
Lysine	170 ± 28	136 ± 47	52 ± 7
Histidine	85 ± 10	103 ± 31	46 ± 3

** $p < 0.01$.

patients were not statistically different from those observed in dialyzed CRF patients before dialysis, the results from these two groups were combined for further analysis. Prior to dialysis essential plasma amino acid levels (valine, leucine, isoleucine, lysine) in CRF patients did not differ significantly from those found in control children, some nonessential amino acid levels (asparagic acid, ornithine) were higher while others (glutaminic acid, proline) were significantly lower, or unchanged (alanine, glycine). This pattern of plasma amino acids is not indicative of protein malnutrition. In response to hemodialysis all plasma amino acids decreased, some of them significantly, as expected.

Urinary excretion of 3-met-HIS as a possible measure of muscle protein breakdown was also assessed. No significant difference was found in 3-met-HIS excretion between healthy children (117.2 ± 38.1 μ*M*/day) and patients with CRF before or after hemodialysis (81.2 ± 43.1 vs. 136.2 ± 118.1 μ*M*/day).

Table III. Urinary excretion of amino acids in healthy children and in children with CRF before and after hemodialysis (μ*M*/day)

	Control	CRF	
		before dialysis	after dialysis
Peptide HYP	225.2±17.7*	33.1±13.5*	43.7±27.1
Free HYP	9.8±3.6**	19.9±6.1**	42.1±21.1
3-met-HIS	117.2±38.1	81.2±43.1	136.2±118,1

* $p < 0.05$; ** $p < 0.01$.

CRF was found to be associated with elevated free HYP (19.9 ± 6.1 vs. 9.8 ± 3.6 μ*M*/day, $p < 0.05$) and depressed peptide HYP excretion (33.1 ± 13.5 vs. 225.2 ± 17.7 μ*M*/day, $p < 0.01$). After hemodialysis free HYP excretion doubled to 42.06 ± 21.1 ($p < 0.05$), whereas peptide-bound HYP remained practically unchanged (43.7 ± 27.1) (table III).

Total GAG excretion rate was significantly lower in CRF patients than in healthy children (7.0 ± 2.4 vs. 16.1 ± 1.9 μ*M* uronic acid/day, $p < 0.05$) and appeared to be unaffected by dialysis therapy (9.0 ± 3.6 μ*M* uronic acid/day).

The distribution of individual GAGs revealed a characteristic pattern. While chondroitin 4-sulfate + chondroitin 6-sulfate (Ch-Ss) were the predominant GAG fractions (65.0 ± 22.1%), in healthy subjects they amounted to only 14.0 ± 9.9% before and 15.1 ± 1.9% after hemodialysis ($p < 0.01$) in CRF. Consequently, the proportion of nonsulfated and under sulfated fractions increased significantly in CRF patients as compared to healthy children (hyaluronic acid (HA) 38.4 ± 12.0 vs. 11.0 ± 3.4%, $p < 0.01$; chondroitin + heparan sulfate (Ch + HS) 36.9 ± 9.5 vs. 20.5 ± 6.2%, $p < 0.05$; keratan sulfate (KS) 5.7 ± 2.0 vs. 4.5 ± 1.8%, $p > 0.05$). Hemodialysis had no significant influence on this pattern of percentage distribution (HA 45.8 ± 16.8%; Ch+HS 27.2 ± 17.8%; KS 10.5 ± 3.7%). Dermatan sulfate (DS) accounted only for a small fraction of about 2% of total GAG excretion and showed no consistent change in CRF patients with or without hemodialysis.

As a result, the respective mean values of daily excretion of Ch-Ss, HA, Ch+HS, KS and DS in healthy children were 10.5 ± 3.4, 1.8 ± 0.6,

3.3 ± 0.9, 0.7 ± 0.2 and 0.4 ± 1.0 as μ*M* uronic acid/day. CRF resulted in an excretion rate of Ch-Ss 1.1 ± 1.0, $p < 0.01$, HA 2.7 ± 1.3, $p < 0.05$, Ch+HS 2.6 ± 1.3, $p > 0.05$, KS 0.4 ± 0.2, $p < 0.01$ and DS 0.2 ± 0.2 μ*M* uronic acid/day, respectively.

Except for HA which increased from 2.7 ± 1.3 to 4.3 ± 1.0 μ*M* uronic acid/day, no consistent changes in the excretion of individual GAG compounds could be observed after hemodialysis.

In an attempt to relate the various parameters of connective tissue metabolism to the severity of CRF, the daily HA excretion was found to correlate positively to plasma creatinine ($r = 0.73$, $p < 0.01$) and plasma urea ($r = 0.71$, $p < 0.05$) and peptide-bound HYP to plasma creatinine ($r = 0.81$, $p < 0.01$).

We failed, however, to demonstrate a significant relationship of the other measured individual GAGs to plasma urea or creatinine. Total or individual GAGs did not correlate with the growth activity of our patients either.

In summary, to assess the characteristics of connective tissue metabolism in CRF, urinary excretion of GAG fractions and HYP was determined in 10 patients with CRF and in 10 age-matched healthy children. CRF was found to be associated with elevated free HYP (19.9 ± 6.1 vs. 9.8 ± 3.6 μ*M*/day, $p < 0.05$) and depressed peptide HYP excretion (33.1 ± 13.5 vs. 225.2 ± 17.7 μ*M*/day, $p < 0.01$), low rate of total GAG excretion (7.0 ± 2.4 vs. 16.1 ± 1.9 μ*M* uronic acid/day, $p < 0.05$) with low Ch-Ss (14.0 ± 9.9 vs. 65.0 ± 22.1%) and high proportion of nonsulfated or undersulfated fractions, i.e. HA+Ch+HS (75.3 ± 11.4 vs. 31.5 ± 5.7%). Urinary 3-met-HIS excretion and plasma essential free amino acids did not differ in the two groups. We conclude that the disturbed collagen metabolism observed in CRF cannot be accounted for by protein malnutrition or enhanced protein breakdown.

References

1 Dull, T.A.; Henneman, P.H.: Urinary hydroxyproline as an index of collagen turnover in bone. New Engl. J. Med. *268:* 132–154 (1963).

2 Prockop, D.J.: Isotopic studies on collagen degradation and the urine excretion of hydroxyproline. J. clin. Invest. *43:* 453–460 (1964).

3 Avioli, L.V.: Collagen metabolism, uremia and bone. Kidney int. *4:* 105–115 (1973).

4 Hallgren, R.; Engström-Laurent, A.; Nisbeth, U.: Circulating hyaluronate. A potential marker of altered metabolism of the connective tissue in uremia. Nephron *46:* 150–154 (1987).
5 Kivirikko, K.J.; Laitinen, O.; Prockop, D.J.: Modifications of a specific assay for hydroxyproline in urine. Analyt. Biochem. *19:* 249–255 (1967).
6 Radha, E.; Bessman, S.P.: A rapid colorimetric method for 3-methylhistidine in urine. Analyt. Biochem. *121:* 170–174 (1982).
7 Tanaka, Y.; Gore, I.: Cellulose column chromatography for the fractionation and isolation of acid mucopolysaccharides. J. Chromat. *23:* 254–260 (1966).
8 Davidson, E.A.: Analysis of sugars found in mucopolysaccharides; in Colowich, Kaplan, Methods in enzymology, vol. VIII, pp. 52–60 (Academic Press, New York 1966).

E. Sulyok, MD, County Children's Hospital, Box-76, H–7601 Pécs (Hungary)

Contr. Nephrol., vol. 67, pp. 86–94 (Karger, Basel 1988)

Alteration in Carnitine Metabolism in Uremic Children

M. Bulla[a], *A. Glöggler*[b], *P. Fürst*[b], *M. Frosch*[a], *E. Kuwertz-Bröking*[a]

[a]University Department of Pediatrics and Nephrology, Münster;
[b]Institute of Biology, Chemistry and Nutrition, Stuttgart-Hohenheim, FRG

In humans, carnitine (a quarternary ammonium base involved in the tissue oxidation of fatty acids, hepatic ketogenesis and the transfer of acetyl-CoA/CoA) is synthesized from lysine and methionine, mainly in the liver and kidney. Therefore, renal insufficiency (RI) is claimed to be accompanied by a severe disturbance in carnitine metabolism [1–3, 10, 13]. The aim of the study was to evaluate in prepubertal children: (1) the alterations in the plasma pattern of carnitine, related to different degrees of progressive RI; (2) the carnitine loss through the dialyzer membrane, as well as the urinary carnitine excretion in children undergoing regular hemodialysis treatment (RDT); in comparison with the concentrations of plasma and urinary carnitine fractions in healthy control children.

Investigation Protocol

(1) To assess the degree of alterations of the carnitine metabolism in relation to the progression of RI, we determined in prepubertal children the plasma total carnitine (pTC), free carnitine (pFC) and its short- and long-chain acyl esters (pSCC and pLCC):

(a) In 7 conservatively treated children, suffering from preterminal progressive renal failure (RI): 4 boys, 3 girls, age: 8.4 ± 5.1 years; weight: 28.4 ± 16.3 kg; height: 125.4 ± 33.0 cm; bone age (Greulich-Pyle): 7.5 ± 3.0 years; pubertal developmental status: Tanner PH 1–2; serum creatinine: 4.28 ± 1.72 mg/dl; endogenous creatinine clearance: 14.12 ± 7.55 ml/min/1.73 m^2 BSA. (b) In 13 uremic children undergoing RDT, both before and after a dialysis session lasting 4 h (bRDT and aRDT): 7 boys, 6 girls, age: 11.1 ± 3.9 years; weight: 39.4 ± 11.8 kg; height: 135.0 ± 20.8 cm; bone age (Greulich-Pyle): 9.8 ± 4.0 years; pubertal developmental status: Tanner PH 1–2; endogenous creatinine clearance: 2.29 ± 2.8 ml/min/1.73 m^2 BSA; time on dialysis: 38.2 ± 17.3 months. In comparison with 20 healthy control children (Co_1) matched for developmental status: 11 boys, 9 girls, age: 6.4 ± 2.3 years; weight: 20.3 ± 1.0 kg; height: 106.7 ± 27.0 cm; serum creatinine: 0.7 ± 0.2 mg/dl.

(2) The carnitine losses through the dialyzer membrane were investigated in 11 prepubertal children undergoing RDT for 23 ± 11 months. Clinical data were as follows:

(a) 8 boys, 3 girls, age: 10.9 ± 5.3 years; weight: 30.0 ± 13.2 kg; height: 1.28 ± 0.38 m; pubertal developmental status: Tanner PH 1–3; endogenous creatinine clearance: 1.8 ± 1.6 ml/min/1.73 m^2 BSA. (b) Of these 11 children, 6 had a residual urinary carnitine excretion: 4 boys, 2 girls, age: 10.3 ± 8.4 years; weight: 30.0 ± 13.2 kg; height: 1.27 ± 0.4 m; pubertal developmental status: Tanner PH 1–3; endogenous creatinine clearance: 3.8 ± 1.4 ml/min/1.73 m^2 BSA; time on dialysis: 11 ± 7 months. (c) The dialysis-induced carnitine losses were compared with the urinary carnitine excretion in 13 healthy children (Co_2) with the following data: 9 boys, 4 girls, age: 9.5 ± 4.3 years; weight: 37.7 ± 14.6 kg; height: 1.4 ± 0.58 m; pubertal developmental status: Tanner PH 1–3; endogenous creatinine clearance: 109.0 ± 10.3 ml/min/1.73 m^2 BSA. All children (RI, RDT, Co_1 and Co_2) were in a stable condition, without any signs of overt infections. Apart from the impairment of the renal function in the RI and RDT children, none suffered from acute or progressive liver/intestinal/metabolic or systemic disease.

Sample Collection and Carnitine Determination

After a 12-hour overnight fast (approximately 36 h after the last dialysis session), EDTA blood samples were taken from RI, RDT and Co_1/Co_2 children under strict body rest.

For determination of carnitine losses in RDT, fasting EDTA blood samples were collected before (bRDT) and 10 min after (aRDT) a hemodialysis session lasting 4 h. These losses were compared with the carnitine appearance rates in dialysate samples (collected after 10, 30, 60, 90, 120, 150 and 180 min, and 10 min before the end of the hemodialysis session), and with the carnitine dialysance (determined in 3 intermittently dialyzed children by repeated collection of EDTA blood samples taken before and after the dialyzer), which was calculated by Fick's principle.

For investigation of the restoration of plasma carnitine levels after dialysis, EDTA blood samples were taken regularly in the interdialytic period in 5 children undergoing RDT. Urinary carnitine excretion was measured in 24-hour urine samples in both RDT children and healthy control children.

After centrifugation at 1,800 *g* for 10 min, plasma samples (1 ml) were frozen at −80 °C until carnitine analysis. Determination of total and free carnitine (pTC, pFC) and its fractions (acylcarnitine = pAC, pLCC, pSCC) in plasma, dialysate and urine was performed by the method of Cederblad and Lindstedt [8] as modified by Rössle et al. [14]. For statistical analysis we used the u test according to Wilcoxon, Whitney and Mann, as well as the Wilcoxon test for paired data and the Student's t test for unselected groups.

Table I. Levels of plasma carnitine (and its fractions) in prepubertal healthy children and in children suffering from progressive renal insufficiency, under conservative treatment (RI) and under RDT

		pFC, μ*M*	pSCC, μ*M*	pLCC, μ*M*	pTC, μ*M*	pAC/pFC
Controls	x̄	38.6	8.0	3.1	49.8	0.31
(n = 20)	± SD	10.9	3.9	0.8	10.0	0.2
RI	x̄	46.0	10.4	2.7*	59.0	0.34
(n = 7)	± SD	26.5	5.1	0.6	31.6	0.1
RDT	x̄	23.9***	24.4***	4.1*	52.4	1.30***
(n = 13)	± SD	9.0	8.2	1.9	15.4	0.5

*** $p < 0.001$, * $p < 0.05$, significant vs. controls.

Table II. Levels of plasma carnitine and its fractions in RDT children before and after a hemodialysis session lasting 4 h, compared with the results in healthy control children (Co_1): mean values ± SD (μmol/l)

	Co_1 (n = 20)	RDT (n = 13)	
		before	after
pFC	33.0 ± 8.5	23.9 ± 9.0**	7.3 ± 3.1***
pSCC	11.2 ± 4.7	24.4 ± 8.2***	6.7 ± 3.9**
pLCC	2.2 ± 0.5	4.1 ± 1.9***	2.2 ± 1.9
pAC/pFC	0.31 ± 0.2	1.30 ± 0.51***	1.40 ± 0.82**

*** $p < 0.001$, ** $p < 0.01$ vs. controls.

Results

(1) *In RI,* pTC and pFC were significantly elevated as compared with the results in Co_1 ($p < 0.01$), while pLCC was markedly lowered ($p < 0.05$) (table I; fig. 1).

In RDT children, bRDT, the mean predialytic pTC level, did not differ from that in the controls (Co_1). However, these apparently normal pTC levels were the result of a significant decrease in pFC ($p < 0.001$) and a

1
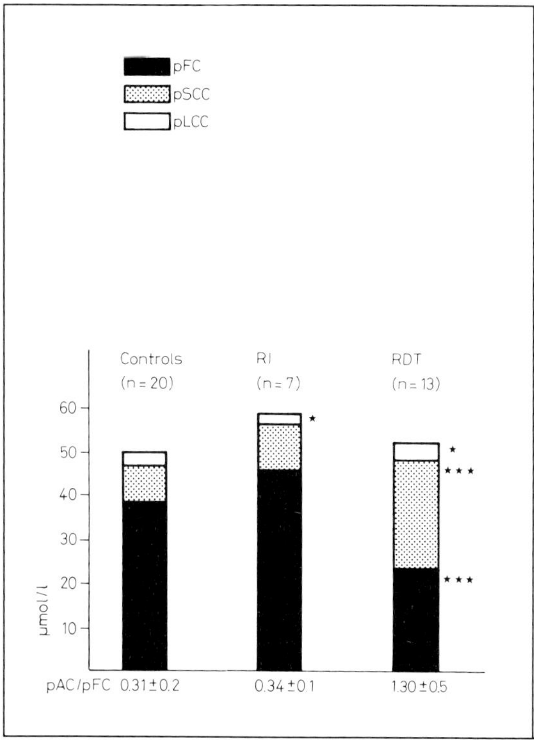

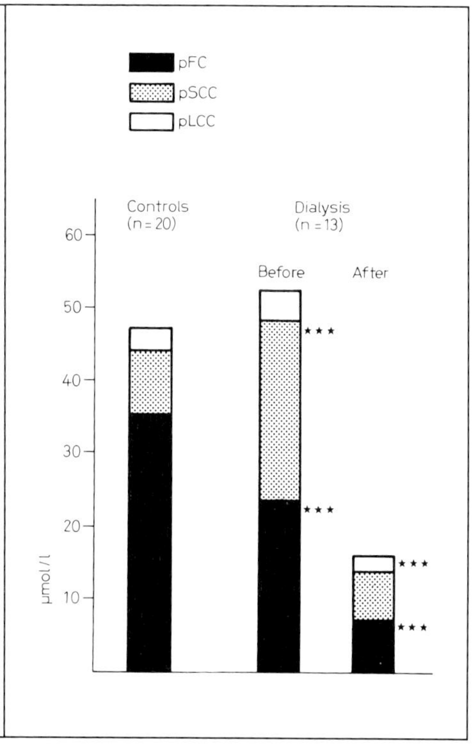

2

Fig. 1. Levels of plasma carnitine (and its fractions) in prepubertal healthy children and in children suffering from progressive renal insufficiency, under conservative treatment (RI) and under RDT. *** $p \leq 0.001$, * $p < 0.05$, significant vs. controls.

Fig. 2. Plasma carnitine fractions in pediatric RDT patients before and after a hemodialysis session lasting 4 h. *** $p \leq 0.001$, significant vs. controls.

corresponding rise in pSCC and pLCC ($p < 0.002$, 0.5), well documented by a pathological value of the pAC/pFC ratio.

As concerns *aRDT,* dialysis treatment reduced all carnitine plasma fractions markedly by a just equal elimination of pFC and pAC: pTC by −69%, pFC by −69%, pSCC by −73% and pLCC by −46%, with $p < 0.001$ for each fraction (fig. 2; table II).

Restoration of the plasma carnitine levels in the interdialytic period was achieved very slowly, in about 36–48 h. We observed a more rapid increase in the first 3–6 h, followed by a delayed, hyperbolic-looking increment of the curve in the following hours (fig. 3).

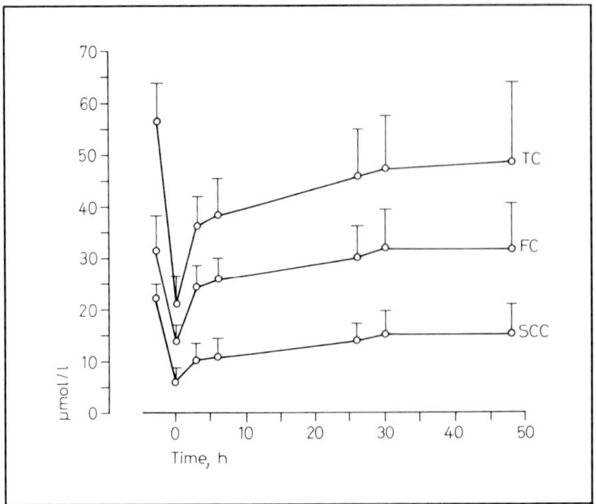
µmol / l
Time, h
TC
FC
SCC

3

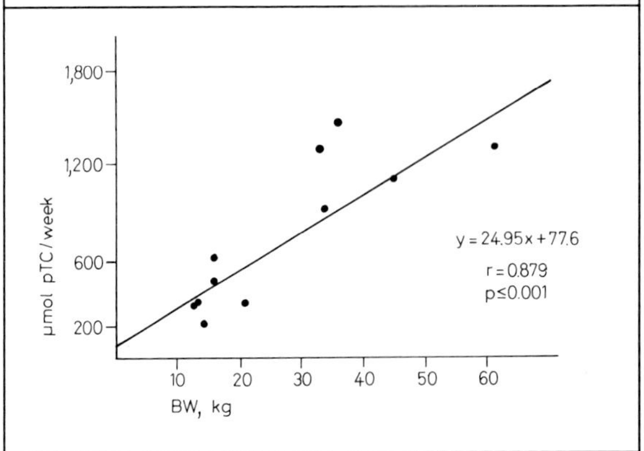
µmol pTC / week
BW, kg
y = 24.95x + 77.6
r = 0.879
p≤0.001

4

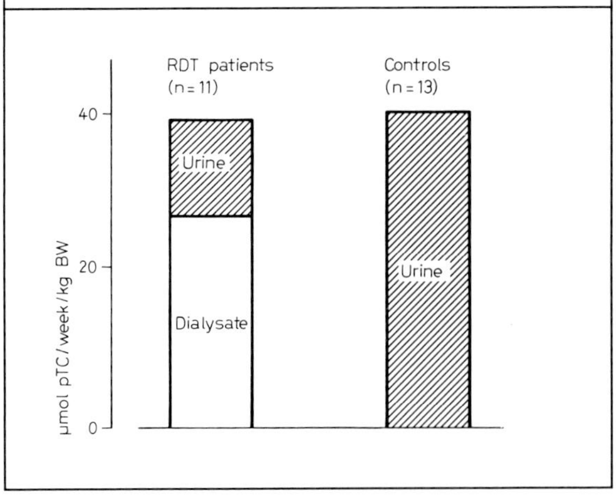
µmol pTC / week / kg BW
RDT patients
(n = 11)
Controls
(n = 13)
Urine
Dialysate
Urine

5

Table III. Total carnitine losses in dialysate through a dialyzer membrane (in a hemodialysis session lasting 4 h) compared with urinary carnitine excretion in healthy control children: mean ± SD (µmol/week/kg BW)

	Carnitine losses in dialysate (n = 11)	Urinary carnitine excretion in RDT children (n = 6)	Σ-urine and dialysate loss	Urinary carnitine excretion in healthy controls (Co_2) (n = 13)
pTC	27.5±9.1	11.88±9.4	39.38	40.50±19.86
pFC	15.4±5.2	4.55±3.0	19.95	21.52±14.05
pSCC	11.9±5.1	6.92±6.4	18.82	17.95±6.81
pLCC	–	0.41±0.3		1.0±0.55
pAC/FC	0.80±0.27	2.04±1.5		1.23±0.69

(2) *The measured TC losses* into the dialysate ranged from 1,278 ± 65 to 915 ± 47.2 µmol/week, with an almost equal distribution for FC and SCC eliminated into the dialysate (FC: 445 ± 30.4 µmol/week; SCC: 315 ± 152 µmol/week). These carnitine losses into the dialysate were directly related to the children's body weight (BW) ($r = 0.84$, $p < 0.001$; fig. 4).

By expressing the weekly carnitine dialysate losses in relation to body weight, we found that the measured carnitine losses in the dialysate of 15.5 ± 5.2 µmol/week/kg BW for FC and 11.5 ± 5.1 µmol/week/kg BW for SCC were well comparable with the *dialysance values* of pFC (19.0 ± 8.1 µmol/week/kg BW) and of pSCC (16.3 ± 5.1 µmol/week/kg BW) determined by Fick's principle in 3 of these RDT children. The *urinary carnitine excretion in 6 RDT children* with residual urinary carnitine output were 4.24 ± 2.8 µmol/week/kg BW for FC and 6.09 ± 5.7 µmol/week/kg BW for SCC. In comparison, the *urinary carnitine excretion in healthy children* amounted to 21.5 ± 14.1 µmol/week/kg BW for FC and 18.0 ± 6.8 µmol/week/kg BW for SCC (table III; fig. 5).

Fig. 3. Restoration of levels of plasma carnitine (and its fractions) in the interdialytic period ($n = 5$, $\bar{x} \pm SD$).

Fig. 4. Total carnitine elimination in relation to body weight.

Fig. 5. Comparison of total carnitine losses in dialysate and urine in regularly dialyzed children with urinary carnitine excretion in healthy children.

Discussion

In accordance with the report by Leschke et al. [12], in *RI children,* we found significantly elevated pTC and pFC levels, ($p < 0.01$), suggesting a disturbed carnitine elimination. As a sign of diminished biosynthesis and disturbed carrier function for long-chain fatty acids, pLCC was significantly lowered ($p < 0.05$). *In RDT children,* the mean predialytic 'normal' levels of pTC were the result of a *significant decrease in pFC and a corresponding increase in pSCC and pLCC,* causing a highly pathological pAC/pFC ratio. These results are well comparable with those in reports on regularly dialyzed adults [6, 9]. This pathological ratio indicated a limited availability of pFC, and hence a higher demand on the carrier system for the transport and turnover of free fatty acids into the mitochondrial matrix. It may be suggested that an inversion of the distribution of enzymes such as carnitine acyltranslocase and/or carnitine acyltransferase may occur, causing a withdrawal of acylcarnitines out of the mitochondria. Therefore, β-oxidation may be reduced, as well as the conversion of acetyl-CoA to free CoA. Disturbances in ketogenesis may also be suspected. In accord with reports in the literature [2, 4, 11, 15], our study revealed that dialysis treatment reduced all pC fractions markedly (–70 to –40%, $p <$ 0.001–0.0005). The measured pTC loss in the dialysate ranged from 1,278 ± 64 to 915 ± 472 μmol/week, depending on the patients' weight ($r =$ 0.84). The bRDT to aRDT relation of the pAC/pFC ratio demonstrated a comparable dialysance of pAC and pFC (pFC 662.2 ± 284, pSCC 588.1 ± 195 μmol/week; pFC 15.5 ± 5.2, pSCC 11.9 ± 5.1 μmol/week/kg BW).

In the literature, the alterations in the carnitine metabolism in RDT patients are mainly explained in terms of the carnitine losses through the dialyzer membrane [2, 4, 15]. In contrast with this idea, we found that the carnitine losses in the dialysate combined with the urinary carnitine excretion were well comparable with the weekly carnitine excretion of healthy control children. However, the slow kinetics of restoration of the predialytic plasma carnitine levels in the interdialytic period indicates an important and long-lasting intracellular imbalance in this period. Our study demonstrated that the plasma carnitine levels are restored extremely slowly in the interdialytic period, in about 36–48 h, with a more rapid increase in the first 3–6 h, followed by a delayed, hyperbolic-looking increase. Since the oral carnitine intake is very low in RDT children (table IV), plasma carnitine correction can occur only *through the delivery of endogenous carnitine from the intracellular pool* (mainly from the muscle). This can induce

Table IV. Exogenous carnitine intake in food in children undergoing regular hemodialysis treatment

Patient No.	Weight, kg	Carnitine intake food		
		μmol/day	mg/day	mg/kg/day
1	16.4	520.7	83.8	5.11
2	17.0	361.4	58.2	3.42
3	20.9	535.6	86.2	4.12
4	20.9	477.6	76.9	3.68
5	20.9	479.1	77.1	3.69
6	24.0	660.3	106.3	4.43
7	34.0	773.6	124.5	3.66
8	36.0	953.6	153.5	4.26
9	36.5	736.5	128.2	3.51
10	41.0	813.0	130.9	3.19
11	50.0	807.5	130.0	2.60
$\bar{x} \pm SD$	28.9 ± 11.1	647.2 ± 183.9	105.1 ± 30.2	3.79 ± 0.67

a severe endogenous carnitine deficiency syndrome, as reported in long-term RDT-treated adults. These data point to the need for further investigations at an intracellular level in muscle and/or other key organs. Elevation of pFC levels by exogenous *L*-carnitine substitution might be of help. Therefore, the carnitine metabolism in plasma and muscle during *L*-carnitine substitution is now being studied.

Conclusion

The altered carnitine metabolism in children suffering from progressive renal insufficiency or undergoing RDT is related to a disturbed de novo biosynthesis and to an altered elimination of carnitine and its fractions through the dialysis membrane. The increased need for pFC (shown by the reduced pFC levels), combined with the decreased biosynthetic capacity, becomes very important in progressive uremia, necessitating regular dialysis treatment. Together with an important intracellular carnitine imbalance in the interdialytic period, due to a delay in restoration of the depleted plasma carnitine levels after RDT, these alterations can cause reduced fatty acid transport, altered ketogenesis and disturbed transfer of acyl-CoA/CoA for energy utilization in the metabolism.

References

1 Bartel, L.L.; Hussey, J.L.; Shargo, E.: Effect of dialysis on serum carnitine, free fatty acids and triglyceride levels in man and rats. Am. J. clin. Nutr. *34:* 1314–1320 (1982).
2 Battistella, P.A.; Angelini, C.; Vergani, L.; Bertoli, M.; Lorenzi, S.: Carnitine deficiency induced during hemodialysis. Lancet *i:* 939 (1978).
3 Böhles, H.: Die Bedeutung des Carnitins für Stoffwechsel und Ernährung. Ernährung *7:* 76–80 (1983).
4 Bøhmer, T.; Begrem, H.; Eiklid, K.: Carnitine deficiency induced during intermittent haemodialysis for renal failure. Lancet *i:* 126–128 (1978).
5 Bøhmer, T.; Ryding, A.; Solberg, A.E.: Carnitine levels in human health and disease. Clinica chim. Acta *57:* 55–61 (1974).
6 Bremer, J.: Carnitine and its role in fatty acids metabolism. Trends Biochem. Sci. *2:* 207–209 (1977).
7 Chan, M.K.; Varghese, Z.; Moorhead, J.F.; Persand, J.W.: Carnitine in haemodialysis patients. Lancet *ii:* 1028–1029 (1980).
8 Cederblad, G.; Lindstedt, S.: A method for the determination of carnitine in the picomole range. Clinica chim. Acta *37:* 235–243 (1972).
9 De Felice, S.L.; Klein, M.I.: Carnitine and hemodialysis. A minireview. Curr. ther. Res. *28:* 195–198 (1980).
10 Gusmano, R.; Oleggini, R.; Perfumo, F.: Plasma carnitine concentrations and dyslipidemia in children on maintenance hemodialysis. J. Pediat. *99:* 429–432 (1981).
11 Leschke, M.; Rumpf, W.G.; Eisenhauer, T.: Quantitative assessment of carnitine loss during hemodialysis and hemofiltration. Kidney int. *24:* suppl. 16, pp. 6143–5146 (1983).
12 Leschke, M.; Rumpf, K.W.; Eisenhauer, Th.; Becker, K.; Bock, U.; Scheller, F.: Serum-Spiegel und Urinausscheidung von *L*-Carnitin bei Patienten mit normaler und eingeschränkter Nierenfunktion. Klin. Wschr. *62:* 274–277 (1984).
13 Mingardi, G.; Bizzi, A.; Cini, M.; Licini, R.; Mecca, G.; Garatini, S.: Carnitine balance in haemodialyzed patients. Clin. Nephrol. *13:* 269–270 (1980).
14 Rössle, C.; Kohse, K.P.; Franz, H.-E.; Fürst, P.: An improved method for the determination of free and esterified carnitine. Clinica chim. Acta *149:* 263–268 (1985).
15 Rumpf, K.W.; Leschke, M.; Eisenhauer, T.; Becker, K.; Köthe, U.; Scheler, F.: Quantitative assessment of carnitine loss during hemodialysis and hemofiltration. Proc. Eur. Dial. Transplant. Ass. *19:* 229–301 (1982).
16 Savica, V.; Bellinghieri, G.; Di Stefano, C.; Corja, E.; Consolo, F.; Corsi, M.; Maccari, F.; Spagnoli, L.G.; Kollaschi, S.; Palmieri, G.: Plasma and muscle carnitine levels in hemodialysis patients with morphological-ultrastructural examination of muscle samples. Nephron *35:* 232–236 (1983).

Prof. M. Bulla, MD, Division of Nephrology, Department of Pediatrics,
University of Münster, Domagkstrasse 3, D–4400 Münster (FRG)

Contr. Nephrol., vol. 67, pp. 95–98 (Karger, Basel 1988)

Albumin Charge in Adriamycin Nephrosis

F. Ginevri, G.M. Ghiggeri, F. Perfumo, G. Candiano, R. Oleggini, R. Bertelli, M.T. Piccardo, R. Gusmano

Nephrology Department, G. Gaslini Institute, Genoa, Italy

Proteinuria in minimal change nephropathy (MCN) and in experimental nephrosis due to aminonucleoside has classically been considered the result of an alteration of fixed anionic sites in the glomerular basement membrane (GBM) [1, 2]. As a consequence, albumin (ALB) which is anionic in charge, is not repelled by glomerular polyanions and is freely lost in urines.

Current concepts on this topic are, however, rapidly on the move and data are now available that point to different, if not conflicting, mechanisms. By this token, charge-mediated renal selectivity towards endogenous ALB is preserved in MCN [3] and in experimental nephrosis no significant alteration of anionic charges of GBM has been documented by various techniques [4].

In this study we used endogenous rat serum ALB (RSA) as a marker of GBM selectivity from rats affected by adriamycin (ADR) nephrosis. In fact, compared to serum ALB from other species, RSA is more cationic and microheterogeneous [5] and therefore a naturally occurring endogenous marker of GBM selectivity.

Material and Methods

Animal Model

Experimental nephrosis was induced in 4 male Sprague-Dawley rats, weighing 300–350 g and fed standard rat chow (21 % casein). ADR was administered as a single dose i.v. (7.5 mg/kg). All animals were studied 14 days after ADR. Six rats given vehicle injections were used as controls.

Albumin Preparation

Serum and urinary ALB were prepared by pseudo-ligand chromatography with Affi Gel Blue as previously described [6] and characterized by immunoelectrophoresis.

IEF and Silver Staining

Ultrathin (240 µm) isoelectric focusing was performed in polyacrylamide gel on silanized glass plates. Polymerization solutions and electrophoresis conditions were as previously described [5]. Silver staining of proteins was performed with the photochemical method of Merril et al. [7].

Fatty Acid Determination

Extracted from ALB, fatty acids (FAs) were determined by isocratic HPLC as 9-anthryldiazomethane fluorescent derivates [8].

Results and Discussion

Compared to serum ALB from several animal species (including humans and bovines in which the pI of ALB is 4.7), RSA is more microheterogenous and more cationic, covering a wide range of pIs from 4.7 to 6.1 [5]. Owing to this unique characteristic, RSA may be considered a useful probe for inspecting the charge-mediated renal properties in ADR nephrosis, in which proteinuria is attributed to the loss of glomerular polyanions [1, 2].

In our experiment rats given ADR in a single dose developed proteinuria within 2 weeks. At this time, the patterns of serum and urinary ALB were markedly different (fig. 1a, b), RSA being less microheterogeneous and more anionic than the urinary homologue. In nephrotic rats urinary ALB was similar to normal RSA [5].

Nephrotic RSA included a main pI 4.7 band and only thin isoforms with a more cationic charge, but in urines all isoalbumins were detectable, the most cationic (pI = 6.1) being also the most represented. In an attempt to characterize these variations in charge, we determined the FA content of each ALB pool and we found that the pool which was most cationic in charge was also the most delipidated. This is in agreement with the knowledge that FAs introduce negative charges on the ALB sequence.

We interpreted the absence of cationic isoalbumins in serum of nephrotic rats as the result of their loss in urines. This implies that the charge-mediated properties of the GBM are preserved in ADR nephrosis, negatively charged molecules being repelled by electrostatic forces. As an

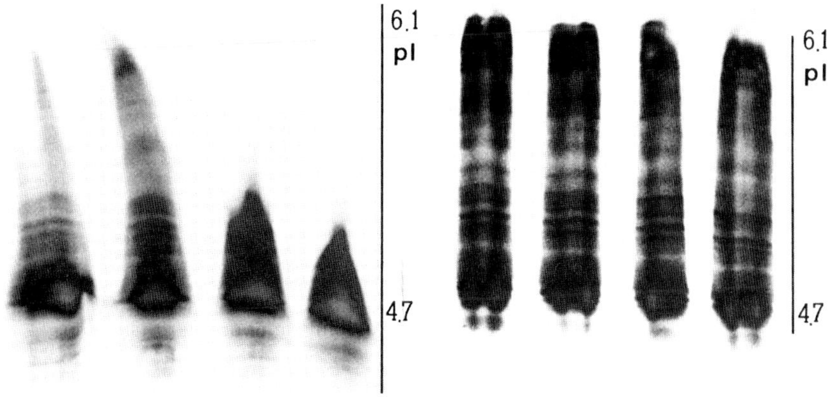

a *b*

Fig. 1. Ultrathin isoelectrofocusing of serum *(a)* and urinary *(b)* ALB from ADR-treated rats. Silver staining. (ADR-casein 21%.)

alternative to this interpretation we can also speculate that the reported difference in pI is produced in situ by delipidating substances which may be present in serum in normal conditions but which are lost in urines in nephrotic syndrome.

These and other possibilities are currently under investigation.

References

1 Carrie, B.J.; Salyer, W.R.; Myers, D.B.: Minimal change nephropathy: an electrochemical disorder of the glomerular membrane. Am. J. Med. *70:* 262–268 (1981).

2 Caulfield, J.P.; Farquhar, M.G.: Loss of anionic sites from glomerular basement membrane in aminonucleoside nephrosis. Lab. Invest. *39:* 505–512 (1978).

3 Ghiggeri, G.M.; Candiano, G.; Ginevri, F.; Gusmano, R.; Ciardi, M.R.; Perfumo, F.; Delfino, G.; Cuniberti, C.; Queirolo, C.: Renal selectivity properties towards endogenous albumin in minimal change nephropathy. Kidney int. *32:* 69–77 (1987).

4 Kanwar, Y.S.; Jakubowski, M.L.: Unaltered anionic sites of glomerular basement membrane in aminonucleoside nephrosis. Kidney int. *25:* 613–618 (1984).

5 Ghiggeri, G.M.; Candiano, G.; Queirolo, C.; Gusmano, R.; Ginevri, F.; Bertani, T.; Remuzzi, G.: Studies by conventional and 'low denaturing' isoelectric focusing on albumin microheterogeneity under normal conditions and in experimental nephrosis. Electrophoresis *8:* 215–220 (1987).

6 Ghiggeri, G.M.; Candiano, G.; Delfino, G.; Queirolo, C.: Highly selective one-step chromatography of serum and urinary albumin on immobilized Cibacron Blue F3 GA. Studies on normal and glycosyl albumin. Clinica chim. Acta *145:* 205–211 (1985).

7 Merril, C.R.; Goldman, D.; Vankeuress, M.L.: Simplified silver protein detection and image enhancement methods in polyacrylamide gels. Electrophoresis *3:* 206–210 (1982).

8 Ghiggeri, G.M.; Candiano, G.; Delfino, G.; Queirolo, C.; Ginevri, F.; Perfumo, F.; Gusmano, R.: Separation of the 9-anthryldiazomethane derivates of fatty acids by high-performance liquid chromatography on a fatty acid analysis column. J. Chromat. *381:* 411–418 (1986).

F. Ginevri, MD, Nephrology Department, G. Gaslini Institute, Via 5 Maggio, 39, I–16148 Genoa (Italy)

Contr. Nephrol., vol. 67, pp. 99–102 (Karger, Basel 1988)

Intraerythrocytary Markers of Peroxidative Alteration in Children with Nephrotic Syndrome

R. Gusmano, F. Ginevri, G.M. Ghiggeri, F. Perfumo, R. Oleggini, G. Candiano, R. Bertelli, M.T. Piccardo

Nephrology Department, G. Gaslini Institute, Genoa, Italy

Despite extensive studies, the mechanism(s) leading to proteinuria in minimal change nephropathy (MCN) is poorly understood. Many insights currently available into the pathogenesis of this disease derive from studies on aminonucleoside nephrosis in rats due to puromycin or adriamycin administration. For more than a decade these models have been extensively investigated in search of an alteration of the charge of glomerular basement membrane (GBM). Up to now there is controversy as to whether in aminonucleoside (AN) nephrosis there is a loss of anionic sites from the GBM [1, 2] or an alteration of the charge-mediated renal selectivity properties [3, 4].

A significant advance in the understanding of the mechanisms of AN toxicity was provided by the recent observation that oxygen free radicals play a role as mediators of the injury. Diamond et al. [5] demonstrated that allopurinol and superoxide dismutase (two potent scavengers of free radicals) were able to protect the kidney from the AN assault resulting in less proteinuria and less severe morphological changes.

Little is known about the possible implication of free radicals in human MCN. Several methodological problems make it difficult to assess free radicals in vivo, especially because most of them have a very short half-life (in the order of a few seconds).

The aim of the present study was to detect indirect signs of cellular peroxidation in children affected by steroid-responsive and unresponsive nephrotic syndrome (SRNS-SUNS). Our hypothesis was that, in the presence of toxic free radicals, cellular signals of lipid peroxidation (such as intracellular malonyldialdehyde – MDA) and of an activation of biological scavengers (reduced glutathione – GSH) should be present. To test this hypothesis we determined intraerythrocytary MDA and GSH levels.

Materials and Methods

Patients

Twenty-nine normal children ranging in age between 3.2 and 14.5 (mean 10.31) and 20 nephrotic children aged between 1 and 13.6 (mean 7.98) were considered. In 10 cases the nephrotic syndrome was responsive to a steroid treatment (SRNS) carried out according the ISKDC schedule [6]. In the remaining 10 children, who were unresponsive to steroids (SUNS), nephrotic syndrome was classified on the basis of renal biopsy as focal glomerulosclerosis (6 patients) and mesangial proliferative glomerulonephritis (4 patients).

In addition, 6 patients (mean age 22.3 years, range 10.5–36) affected by nephrotic syndrome due to various glomerular diseases (2 diabetic nephropathies, 1 membranous nephropathy, 2 lupus nephritis, 1 IgA nephropathy) were also evaluated.

Intraerythrocytary MDA and GSH

Aliquots of erythrocytes suspension (5% v/v) were treated with 0.5 vol of 28% w/v trichloracetic acid. For MDA determination 1 vol of 0.1 *M* thiobarbituric acid in 1 *M* NaOH was added to 4 vol of the acid supernatant and the mixture heated in a boiling water bath for 15 min. Samples were then cooled and absorbance measured at 532 nm [7]. GSH was determined as acid-soluble sulfhydryl groups in the same trichloracetic supernatant. Sulfhydryl content was determined by adding 1 vol of supernatant to 2 vol of 1 *M* Tris (pH 8.0), adding 0.2 vol of 0.01 *M* 5,5-dithiobis(2-nitrobenzoic acid) (pH 7.0) and measuring absorbance at 412 nm [7].

Table I. Intraerythrocytary content of MDA and reduced GSH in children affected by SRNS and SUNS

Patients	n	MDA, n*M*/mg Hb	GSH, n*M*/mg Hb
Normals	29	0.090 ± 0.01	8.39 ± 0.34
SRNS	10	0.103 ± 0.02	6.52 ± 0.51**
SUNS	10	0.213 ± 0.05**	5.10 ± 0.43***
Other nephrotic syndromes	6	0.106 ± 0.02	7.98 ± 0.89

** $p < 0.01$; *** $p < 0.001$, compared to normals. Results are given as mean ± SEM.

Results

In table I the intraerythrocytary levels of MDA and GSH in children affected by SRNS and SUNS are reported. While in SRNS, the MDA concentration was not different compared to normal children, in SUNS it was statistically higher. In both groups of children the content of GSH was significantly lower compared to normal children. No variation in either parameter was found in the group of patients with other nephrotic syndromes.

Discussion

Oxygen free radicals have recently been proposed as mediators of renal injury in AN nephrosis [5]. From this standpoint we hypothesized their participation in the pathogenesis of idiopathic nephrotic syndrome. Since most free radicals have a very short life (in the order of a few seconds) and several technical problems make their quantitation in biological fluids difficult, we took the presence of the effects of a peroxidative assault on biological systems as an indirect proof of their presence.

Erythrocytes offered a very good model for this study since they are targets of free radicals toxicity in vivo and in vitro, and changes due to their peroxidation have been extensively documented [8, 9]. These changes include variations of the spatial arrangement and aggregation of spectrin [8], alterations of the inner layer ghost phospholipids [9] which result in an increase of the intracellular levels of MDA, the major product of lipid breakdown [8]. Finally, the concentration of the intraerythrocytary natural scavengers of free radicals such as glutathione (in the case of H_2O_2) and hemoglobin (in the case of O_2) becomes abnormally low in an attempt to scavenge the toxic compounds [7].

The data we report here on the intraerythrocytary levels of MDA and GSH support the concept of a participation of free radicals in the idiopathic nephrotic syndrome we studied. We were unable to find the same pattern in patients with nephrotic syndromes due to other nephropathies.

We therefore hypothesize that in idiopathic nephrotic syndrome in children free radicals could be considered to be mediators of renal injury, as recently documented in several renal alterations [10–13].

Our next step is to approach the problem with more appropriate techniques.

In summary, intraerythrocytary levels of MDA and reduced GSH taken as markers of peroxidation were determined in children affected by SRNS and SUNS. From the results, the involvement of free radicals can be hypothesized.

References

1 Caulfield, J.P.; Farquhar, M.G.: Loss of anionic sites from glomerular basement membrane in aminonucleoside nephrosis. Lab. Invest. *39:* 505–512 (1978).

2 Lelongt, B.; Makino, H.; Kanwar, Y.S.: Status of glomerular proteoglycans in aminonucleoside nephrosis. Kidney int. *31:* 1299–1310 (1987).

3 Bertolatus, J.A.; Hunsicker, L.G.: Glomerular sieving of anionic and neutral bovine albumins in proteinuric rats. Kidney int. *28:* 467–476 (1985).

4 Ghiggeri, G.M.; Candiano, G.; Queirolo, C.; Gusmano, R.; Ginevri, F.; Bertani, T.; Remuzzi, G.: Studies by conventional and 'low denaturing' isoelectric focusing on albumin microheterogeneity under normal conditions and in experimental nephrosis. Electrophoresis *8:* 215–220 (1987).

5 Diamond, J.R.; Bonventre, J.V.; Karnosky, M.J.: A role for oxygen free radicals in aminonucleoside nephrosis. Kidney int. *29:* 478–483 (1986).

6 International Study of Kidney Disease in Children: The nephrotic syndrome in children. Prediction of histopathology from clinical and laboratory characteristics at the time of diagnosis. Kidney int. *13:* 160–166 (1978).

7 Dise, L.A.; Goodman, D.B.P.: t-Butyl hydroperoxide alters fatty acid incorporation into erythrocyte membrane phospholipid. Biochim. biophys. Acta *859:* 69–78 (1986).

8 Snyder, L.M.; Fortier, N.L.; Trainor, G.; Jacobs, J.; Leb, L.; Lubin, B.; Chiu, D.; Shohet, S.; Mohandas, N.: Effect of hydrogen peroxide exposure on normal human erythrocyte deformability, morphology, surface characteristics and spectrin-hemoglobin cross-linking. J. clin. Invest. *76:* 1971–1977 (1985).

9 Jain, S.K.: In vivo externalization of phosphatidylserine and phosphatidylethanolamine in the membrane bilayer and hypercoagulability by the lipid peroxidation of erythrocytes in rats. J. clin. Invest. *76:* 281–286 (1985).

10 Johnson, R.J.; Couser, W.G.; Chi, E.Y.; Adler, S.; Klebanoff, S.J.: New mechanism for glomerular injury – myeloperoxidase-hydrogen peroxide-halide system. J. clin. Invest. *79:* 1379–1387 (1987).

11 Boyce, N.W.; Holdsworth, S.R.: Hydroxyl radical mediation of immune renal injury by desferrioxamine. Kidney int. *30:* 813–817 (1986).

12 Rehan, A.; Johnson, K.J.; Wiggins, R.G.; Kunkel, R.G.; Ward, P.A.: Evidence for the role of oxygen radicals in acute nephrotoxic nephrites. Lab. Invest. *51:* 396–403 (1984).

13 Rehan, A.; Johnson, K.J.; Kunkel, R.G.; Wiggins, R.C.: Role of oxygen radicals in phorbol myristate acetate-induced glomerular injury. Kidney int. *27:* 503–511 (1985).

Prof. R. Gusmano, Nephrology Department, G. Gaslini Institute, Via 5 Maggio, 39, I-16148 Genoa (Italy)

Contr. Nephrol., vol. 67, pp. 103–106 (Karger, Basel 1988)

Tryptophan Availability in Nephrotic Syndrome

J. Fydryk, W. Szuba, O. Kurzawska, G. Zielazkowska
Institute of Paediatrics, Szczecin, Poland

In an active nephrotic syndrome (NS) the level of both total serum tryptophan (TRP) and its bound-to-albumin fraction are significantly decreased [4]. Availability of TRP for protein and serotonin synthesis depends on its serum level and also on the ratio of TRP to the other essential amino acids sharing the same transport system through cell membranes and blood-brain barrier [6].

The aim of this study was to examine the serum level of essential amino acids which are competitively transported by the same transport system and to estimate the molar ratio of TRP to the remaining amino acids of this group in nephrotic children at various periods of the activity of NS.

Material and Methods

The study covered 33 children aged 2.25–14 years with steroid-responsive NS, lasting 6 months to 13 years (mean 3 years). According to ISKDC [5] they were classified as cortico-dependent (CD, 13 patients), frequent relapsers (FR, 10 patients) and infrequent relapsers (IR, 10 patients). Serum level of TRP, Met, Tyr, Phe, Val, Ile, Leu and albumin was determined 3 times in the course of NS: (I) in all patients in full relapse, before prednisone administration: (II) at the end of the 4th week of treatment with prednisone, after achieving full remission in IR and FR and partial remission in CD; (III) only in IR and FR who remained in remission 4 weeks or more without prednisone, and in 8 CD cases who after cytostatic treatment also fulfilled this criterion.

Total serum TRP concentration (tTRP) and its free fraction (fTRP) after filtration on Amicon CF-25 filters were determined by the fluorometric method [2]. Bound-to-albumin fraction of TRP (bTRP) was calculated by subtracting the free fraction from total serum TRP concentration. The other amino acids were determined on an automatic amino acid analyzer, and serum albumin with the use of immunoplates (M-Partigen, Behringwerke).

Statistical analysis of the data was performed using Student's t test.

Table I. Serum level of amino acids in relapse (I), after 4 weeks of prednisone (II) and in remission without prednisone treatment (III)

Amino acid	Controls	NS		
		I	II	III
tTRP	74±9[1]	29±13*	46±12	52±16
bTRP	66±6	21±13**	40±13	46±16
fTRP	7±5	8±3	5±2	5±1
Met	17±6	19±6	17±5	18±4
Tyr	53±18	57±17	57±9	59±13
Phe	43±14	40±11	40±10	42±13
Val	166±37	176±37	161±28	170±23
Ile	51±13	54±17	44±10	45±10
Leu	98±14	96±30	83±21	88±18

* $0.005 < p$ control/I < 0.01; ** $0.001 < p$ control/I < 0.005.
[1] μmol/l; mean ± 1 SD.

Results

Only in full relapse of NS were tTRP and bTRP significantly decreased, the other amino acids were in normal concentration. After 4 weeks of prednisone and at the treatment-free period, all amino acids remained normal (table I). A good correlation was noted between serum TRP (total and bound) and the serum albumin level (table II). No difference between CD/FR and IR in respect to the TRP level was found (table III). The molar ratio of tTRP to both branched-chain and all measured essential amino acids was low in relapse and increased progressively at subsequent points of estimation, the greatest difference being noted between relapse (I) and the end of the 4-week prednisone treatment (II) (table IV).

Comment

Since about 80% of tTRP circulates in blood in the bound-to-albumin form, a decreased level of tTRP in a relapse of NS can be related to concomitant hypoalbuminemia. According to published data and our own

Table II. Correlation coefficient (r) between serum albumin concentration and tTRP/bTRP

	I	II	III
Serum albumin, g/l	12±7	34±16	47±5
r: Albumin-tTRP	0.412	0.726	0.908
r: Albumin-bTRP	0.569	0.824	1.00

Table III. Serum level of tTRP and bTRP in CD/FR and IR nephrotic children

	I		II		III	
	tTRP	bTRP	tTRP	bTRP	tTRP	bTRP
CD/FR	31±15[1]	23±15	46±11	39±13	59±14	54±13
IR	23±4	15±6	42±13	37±14	35±14	28±8
p	n.s.		n.s.		n.s.	

n.s. = Not significant.
[1] μmol/1; mean ± 1 SD.

Table IV. Molar ratio of serum tTRP to branched-chain and all measured essential amino acids

	tTRP / Val + Ile + Leu	tTRP / Met + Tyr + Phe + Val + Ile + Leu
I	0.089	0.065
II	0.161	0.115
III	0.173	0.124

observations, the reduced binding of TRP to albumin and, consequently, the low level of bTRP result from the elevated ratio of NEFA to serum albumin, where nonesterified fatty acids act as a ligand competing with TRP for binding sites on the albumin molecule [1, 4, 6].

The normal level of the other measured amino acids is not clear. It can be related to the high protein intake, efficient tubular reabsorption or some other factors.

Important implications stem from the ratio of TRP to the remaining essential amino acids. A group of authors from MIT, Boston, Mass. [3], have shown that the low ratio of tTRP to the other amino acids sharing the same transport system hampers TRP entry into the cells and through the blood-brain barrier. Our own results indicate a close relation between the active stage of NS and the decreased TRP molar ratio. That may entail a reduced intracellular concentration of TRP with obvious consequences, and can be important in resistant long-lasting forms of NS.

Conclusions

In an active NS, a low serum level of TRP and a normal level of the other essential amino acids sharing the same transport system through the cell membranes and blood-brain barrier were found. A low molar ratio of TRP to the mentioned amino acids creates unfavourable conditions for the transport of TRP to the intracellular space, and thereby potentially decreases the availability of this amino acid for biosynthesis of protein and serotonin. This may be important in NS which are resistant to treatment.

References

1 Curzon, G.; Knott, P.J.: Environmental, toxicological and related aspects of tryptophan metabolism with particular reference to the central nervous system. CRC crit. Rev. Toxicol. *5:* 145–187 (1977).

2 Denckla, W.D.; Dewey, H.K.: The determination of tryptophan in plasma, liver and urine. J. Lab. clin. Med. *69:* 160–169 (1967).

3 Fernstrom, J.W.; Wurtman, R.J.; Hammerstrom-Wiklund, B.; Rand, W.M.; Munro, H.N.; Davidson, Ch.S.: Diurnal variations in plasma concentrations of tryptophan, tyrosine and other neutral amino acids: effect of dietary protein intake. Am. J. clin. Nutr. *32:* 1912–1922 (1979).

4 Fydryk, J.; Kurzawska, O.; Szuba, W.: Serum tryptophan level and its binding capacity to albumin in nephrotic syndrome. Int. J. pediat. Nephrol. *5:* 124 (1984).

5 International Study of Kidney Disease in Children: Nephrotic syndrome in children: a randomized trial comparing two prednisone regimens in steroid-sensitive patients who relapse early. J. Pediat. *95:* 239–243 (1979).

6 Madras, B.K.; Cohen, E.L.; Messing, R.; Munro, H.N.; Wurtman, R.J.: Relevance of free tryptophan in serum to tissue tryptophan concentrations. Metabolism *23:* 1107–1116 (1974).

J. Fydryk, MD, Institute of Paediatrics, ul. Unii Lubelskiej 1,
71–344 Szczecin (Poland)

Contr. Nephrol., vol. 67, pp. 107–110 (Karger, Basel 1988)

Tonsillar Distribution of IgA, IgG Immunocytes and IgA Subclasses in IgA Nephropathy

J. Nagy[a], *P. Brandtzaeg*[b]

[a] Laboratory for Immunopathology, Second Department of Medicine, University Medical School, Pécs, Hungary;
[b] Laboratory for Immunohistochemistry and Immunopathology, Institute of Pathology, University of Oslo, Rikshospitalet, Oslo, Norway

IgA nephropathy (IgA NP), also termed Berger's disease, is an IgA-containing immune complex-mediated chronic glomerulonephritis (GN) with a large amount of IgA (mainly IgA1) in the mesangium of the glomeruli. The origin of circulating and deposited IgA is unknown. The naso- and oropharynx are common sites of contact with exogenous antigens and the presence of a large number of immunocytes in the tonsils suggests that they have a local immune function. It is known that, at the onset or in the course of IgA NP, upper respiratory tract infections, especially tonsillitis, are frequently followed by periods of gross hematuria. Most patients have raised levels of IgA dimers or larger polymers in their sera [6] and pharyngeal washings [7] as well as a significantly increased percentage of IgA-dimer-producing cells in their tonsils [1, 4], suggesting a tonsillar contribution of IgA in IgA NP.

The aim of this work was to study the possible pathogenetic relationship between tonsillitis and IgA NP, examining the possible abnormalities in the tonsillar IgA-cell system.

Material and Methods

Fourteen tonsils of 7 patients with IgA NP, 1 woman and 6 men, aged between 27 and 45 years were examined. Tonsillectomy was indicated because of recurrent tonsillitis that apparently caused acute exacerbation of renal disease in most of the patients. The *control* group indicated 16 tonsils removed from 8 subjects because of numerous attacks of acute tonsillitis. The controls were not suffering from any renal disease.

Direct paired immunofluorescence staining was used with different fluorescein isothiocyanate (green) and tetramethylrhodamine isothiocyanate (red) conjugates on serial tissue sections [2, 3] for the examination of the relative distribution of: (a) IgA- and IgG-producing cells; (b) dimer (J-chain positive) and monomer (J-chain negative) IgA-producing cells, and (c) IgA1- and IgA2-producing cells.

Cells showing green and red fluorescence were counted in a systemic manner in the extrafollicular area of the tonsils. Immunocyte counts were expressed as medians of the percentages of class ratios within the 2 tonsils of each patient. Group (patients and controls) comparisons were made by the Mann-Whitney nonparametric test (one-tailed).

Results

Table I shows that there was a significant increase in the percentage of IgA-producing cells in the tonsils of patients with IgA NP compared with controls.

On the basis of the presence or absence of J-chain expressed by IgA immunocytes, the ratio of dimer- (J-chain positive) to monomer-producing cells was determined. Significantly more of the former category were found in NP patients than in controls (table II).

A scatter diagram of subclass percentages of IgA-producing cells calculated for patients and controls demonstrated a striking predominance of IgA1 immunocytes in both groups, and the proportion of this subclass tended to be higher in the NP patients (fig. 1).

Discussion

The present and previous [1, 4] studies showed that chronically inflamed tonsils from patients with IgA NP have a higher percentage of IgA-producing cells than controls suffering from chronic tonsillitis. Moreover, the proportion of tonsillar IgA cells expressing J-chains is raised in IgA NP. These disparities may reflect dissimilar antigenic stimuli or genetically determined immunoregulatory differences. Whether the IgA1 dimer-producing cells are largely generated in the tonsils or become terminally differentiated thereafter migrating into these organs remains unknown. They could well have been generated after antigenic stimulation in other mucosal sites of the upper respiratory tract where the same immunocyte isotype predominates [5]. Nevertheless, our findings are compatible with the hypothesis that mesangial dimeric IgA1 is derived from the upper respiratory tract and/or tonsils.

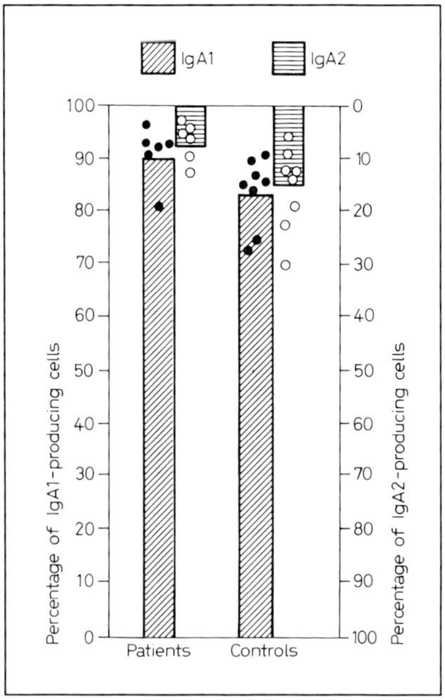

Fig. 1. Scatter diagram of subclass percentages of IgA-producing cells in the tonsils of patients with IgA NP and controls.

Table I. IgG- and IgA-producing cells in the tonsils of IgA NP patients and controls

	IgA NP, %	Controls, %	p
IgG	51.5	63.2	<0.001
IgA	48.5	36.8	<0.001

Table II. Polymer (J-chain positive) and monomer (J-chain negative) IgA-producing cells in the tonsils of IgA NP patients and controls

	IgA NP, %	Controls, %	p
Dimer IgA	35.6	18.8	<0.001
Monomer IgA	64.4	81.2	<0.001

Conclusion

The tonsils of 7 IgA NP patients contained a significantly increased proportion of IgA-producing cells, dimeric IgA-producing cells and IgA1-producing cells. These results are compatible with the hypothesis that in some patients with IgA NP, the IgA deposited in the mesangium may be of tonsillar origin.

References

1 Bene, M.; Faure, G.; Hurault de Ligny, B.; Kressler, M.; Duheille, J.: Immunoglobulin A nephropathy. Quantitative immunohistomorphometry of the tonsillar plasma cells evidences in inversion of the immunoglobulin A versus immunoglobulin G secreting cell balance. J. clin. Invest. *71:* 1342–1347 (1983).

2 Brandtzaeg, P.; Rognum, T.: Evaluation of tissue preparation methods and paired immunofluorescence staining for immunocytochemistry of lymphomas. Histochem. J. *15:* 655–689 (1983).

3 Brandtzaeg, P.; Surjan, L.; Berdal, P.: Immunoglobulin systems of human tonsils. I. Control subjects of various ages: Quantification of Ig-producing cells, tonsillar morphometry and serum Ig concentrations. Clin. exp. Immunol. *31:* 367–381 (1978).

4 Egido, J.; Blasco, R.; Lozano, L.; Sancho, J.; Garcia-Hoyo, R.: Immunological abnormalities in the tonsils of patients with IgA nephropathy: inversion in the ratio of IgA:IgG bearing lymphocytes and increased polymeric IgA synthesis. Clin. exp. Immunol. *57:* 101–106 (1984).

5 Kett, K.; Brandtzaeg, P.; Radl, J.; Haaijman, J.: Different subclass distribution of IgA-producing cells in human lymphoid organs and various secretory tissues. J. Immun. *136:* 3651–3655 (1986).

6 Lopez-Trascasa, M.; Egido, J.; Sancho, J.; Hernandez, L.: IgA glomerulonephritis Berger's disease: evidence of high serum levels of polymeric IgA. Clin. exp. Immunol. *42:* 247–253 (1980).

7 Tomino, Y.; Endoh, M.; Kaneshige, H.; Nomoto, Y.; Sakai, H.: Increase of IgA pharyngeal washings from patients with IgA nephropathy. Am. J. med. Sci. *286:* 15–18 (1983).

Judit Nagy, MD, Laboratory for Immunopathology, Second Department of Medicine, University Medical School, 5 Széchenyi Squ., H–7621 Pécs (Hungary)

Contr. Nephrol., vol. 67, pp. 111–116 (Karger, Basel 1988)

Experimental Glomerulonephritis Induced by the Glomerular Deposition of IgA-Concanavalin A Complexes

J.-C. Davin[a], *J. Nagy*[b], *J. Lombet*[a], *J.-B. Foidart*[a], *C. Dechenne*[a], *P.R. Mahieu*[a]

[a] Departments of Pediatrics and Medicine, State University of Liège, Belgium;
[b] Second Department of Medicine, University Medical School, Pécs, Hungary

It is generally admitted that both Berger's disease and Henoch-Schönlein purpura nephritis (IgA-associated nephropathies) result from the glomerular deposition of circulating immune complexes [1–4].

In addition, numerous clinical observations indicate that acute episodes of macroscopic hematuria occurring in those nephropathies are frequently accompanied, or shortly preceded, by respiratory or gastrointestinal tract infections. Some of those infectious agents (viruses, certain *Escherichia coli* serotypes, for example) bear lectin-like proteins on their surface [5, 6]. The IgA1 subtype of IgA molecules contains both N-glycosidically and o-glycosidically linked oligosaccharides [7]. The possibility therefore exists that circulating IgA1-lectin complexes could appear in the bloodstream during those bouts of hematuria and before the immune response becomes operative. Indeed, we have already shown that some circulating IgA immune complexes can be dissociated by 10-m*M* solutions of *D*-galactose or of α-methyl-mannoside [8].

In the present work, we will provide supportive evidence that circulating IgA-lectin complexes can deposit into rat glomeruli and trigger an acute inflammatory reaction. To that purpose, we have prepared complexes of IgA and concanavalin A (Con A; a lectin recognizing α-*D*-mannosyl residues) for the following reasons: (1) Golbus and Wilson [9] have already demonstrated that Con A is capable of binding to the glomerular capillary wall in rats; (2) autochthonous IgG antibodies bearing terminal mannosyl residues preferentially accumulate into rabbit kidneys [10]; (3) a mannan-binding protein has recently been purified from normal rabbit serum [11];

(4) common environmental antigens implicated in IgA-associated nephropathies, such as gluten or *E. coli* [4], contain mannose-recognizing lectins [5, 12].

Materials and Methods

Preparation of IgA-Lectin Complexes

The binding of human secretory or monoclonal IgA (sIgA and mIgA) to Con A in vitro and its specific inhibition by mannose was preambly ascertained by radioimmunoassay [13]. IgA-Con A complexes were obtained by mixing, in 1 ml of phosphate-buffered saline, pH 7.4 (PBS) and containing 2 m*M* $CaCl_2$, 500 µg of Con A labelled with rhodamine (Con A R; Nordic, Tilburg, The Netherlands), with 500 µg of sIgA (Sigma, St. Louis, Mo.) or of mIgA (Calbiochem, Brussels, Belgium), for 1 h at 37 °C. At the end of incubation, complexes exhibiting an apparent molecular weight of more than 1,000 kdaltons were separated from free IgA or free Con A by gel filtration chromatography on Ultrogel AcA 22 Column (LKB, Bromma, Sweden) [13].

Experimental Protocol

Female Sprague-Dawley rats ($\approx$200 g) were used for all experiments. Rats were anesthetized with Inactin® (BYK, Konstanz, FRG) and the aorta cannulated. The tip of the catheter was located just above the renal arteries. Groups of 3 rats were injected with 0.5 ml of PBS containing 500 µg of either Con A R, IgA, or Con A R-IgA complexes (followed or not by the injection of 1 ml of 1 *M* α-methyl-mannoside in isoosmotic solution). A control group was injected with 0.5 ml of PBS only. Rats were sacrificed 1 h later and kidneys were excised for morphological studies.

Morphological Studies

For light microscopy, portions of the kidneys were fixed in Bouin's renal fixative, sectioned at 4 µm and stained with hematoxylin and eosin, periodic acid-Schiff or Masson's trichrome reagents. For electron microscopy, tissues were fixed by immersion fixation in Karnovsky's fixative, thin sections were stained with uranyl acetate and lead citrate, and they were examined in a Philips EM 201 microscope. For immunofluorescence studies, kidneys were snap frozen in liquid nitrogen. Kidney sections were first incubated with a rabbit antihuman α-chain antiserum (Behring-Werke AG, Marburg, FRG) and then stained either with fluorescein isothiocyanate (FITC)-conjugated goat antirabbit IgG antibodies or with FITC-conjugated rabbit antirat IgG, rat C3 and rat fibrinogen/fibrin antibodies (Nordic).

Results

Immunofluorescence Studies

The results are summarized in table I. It should be noted that, in rats injected with Con A R-IgA complexes, most glomeruli contained granular deposits of Con A, human IgA and rat C3. The specificity was ascertained

Table I. Immunofluorescence microscopy data in the various groups of rats

Group of rats	Glomerular deposits[1]				
	lectin	human IgA	rat C3	rat fibrinogen	rat IgG
Con A-IgA	+++ (granular)	+++	++	++	+
Con A-IgA + α-methylmannoside	+	+	±	±	±
Con A alone	+ (nearly continuous)	–	–	–	–
IgA alone	–	–	–	–	–
PBS	–	–	–	–	–

[1] Intensity of fluorescence was arbitrarily estimated using a scale from 0 to 3+. Rats were sacrificed 1 h after the various intraaortic injections.

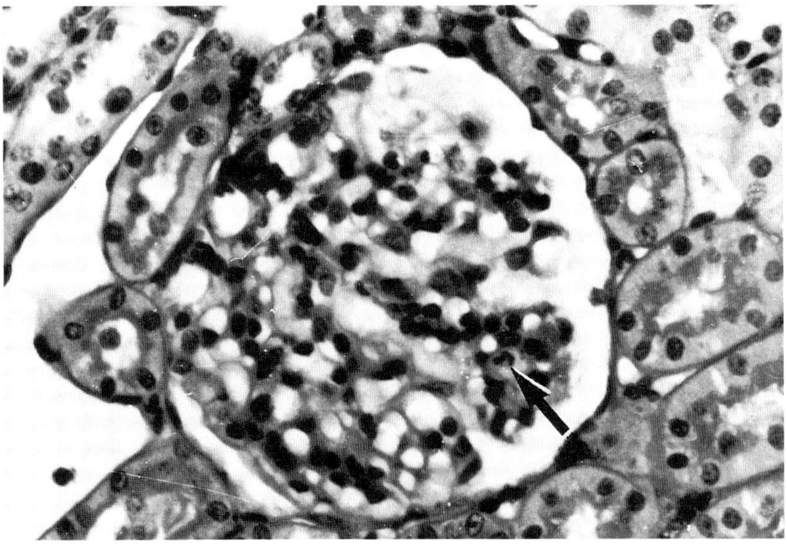

Fig. 1. Note the segmental hypercellular aspect, the hypertrophy of mesangial areas and the presence of a polymorphonuclear leukocyte (arrow). PAS staining. ×400.

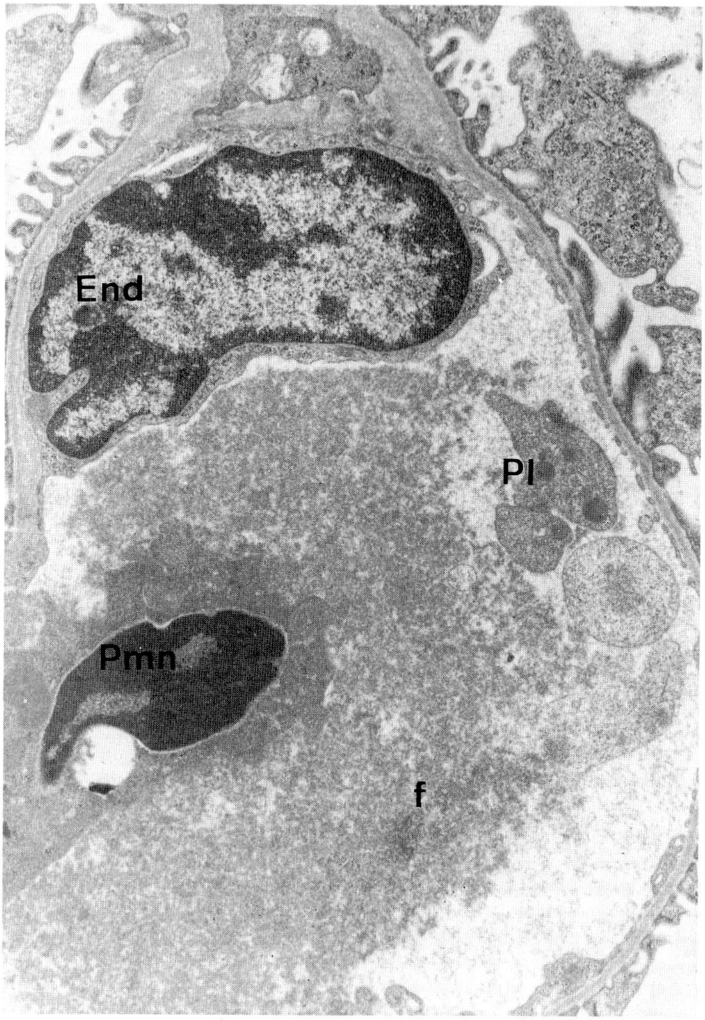

Fig. 2. Clusters of fibrinoid material (F) and aggregated platelets (Pl) obstruct a capillary lumen. A degranulated polymorphonuclear leukocyte (Pmn) is also observed in close contact with the capillary wall. End = Endoplasmic reticulum. Ur.Pb. × 12,000.

by the inhibition of glomerular deposition of IgA-Con A complexes by the subsequent injection of 1 ml of 1 *M* α-methylmannoside in isoosmotic solution.

Light Microscopy

All glomeruli exhibited a turgescent aspect. The mesangial areas appeared to be enlarged and more cellular than in the control groups (fig. 1). Some polymorphonuclear leukocytes (fig. 1) infiltrated the capillary loops in which clusters of fibrinoid material were observed segmentally.

Electron Microscopy

Aggregates of platelets, fibrinoid material and degranulated polymorphonuclear leukocytes were observed in some capillary lumens (fig. 2). In other areas, this fibrinoid material was phagocytosed by intraluminal mononuclear cells in which lysosomial vacuoles were obvious (not illustrated). Numerous mesangial cells exhibited signs of hyperactivity, such as an increased number of polyribosomes, a well-developed rough endoplasmic reticulum and an overload of basement membrane-like material. The presence of red blood cells was noticed in some tubular lumens (data not shown).

Conclusions

The intraaortic injection of IgA-Con A complexes determine, in rats, segmental glomerular lesions similar to those observed in some IgA-associated nephropathies.

References

1 Berger, J.; Hinglais, N.: Les dépôts intercapillaires d'IgA-IgG. J. Urol. Néphrol. *74:* 694–695 (1968).
2 Levy, M.; Broyer, M.; Arsan, A.; Levy-Bontolila, D.; Habib, R.: Glomérulonéphrites du purpura rheumatoïde chez l'enfant. Histoire naturelle et étude immunopathologique; dans Hamburger, Crosnier, Funck-Brentano, Actualités néphrologiques, p. 174 (Flammarion Médecine-Sciences, Paris 1976).
3 Garcia-Fuentes, M.; Chantler, C.; Williams, D.G.: Cryoglobulinemia in Henoch-Schönlein purpura. Br. med. J. *ii:* 163–165 (1977).
4 Woodroffe, A.J.; Gormly, A.A.; McKenzie, P.E.; Wootton, A.M.; Tompson, A.J.; Seymour, A.E.; Clarkson, A.R.: Immunologic studies in IgA nephropathy. Kidney int. *18:* 366–374 (1980).

5 Firon, N.; Ofek, J.; Sharon, M.: Interaction of mannose-containing oligosaccharides with the fimbriae lectin of *Escherichia coli.* Biochem. biophys. Res. Commun. *105:* 1426–1431 (1982).
6 Meager, A.; Hughes, R.C.: Virus receptors; in Cuatrecasas, Greaves, Receptors and recognition, ser. A, vol. 4, p. 143 (Chapman & Hall, London 1977).
7 Baenziger, J.; Kornfeld, S.: Structure of the carbohydrate units of IgA1 immunoglobulin. II. Structure of the O-glycosidically linked oligosaccharide units. J. biol. Chem. *22:* 7270–7281 (1974).
8 Davin, J.C.; Malaise, M.; Foidart, J.B.; Mahieu, P.: Evidence that carbohydrates of the Fc domain of IgA are involved in the formation of large IgA-immune complexes in IgA-associated nephropathies. Xth Int. Congr. of Nephrology, London 1987, book of abstracts, p. 322.
9 Golbus, S.M.; Wilson, C.B.: Experimental glomerulonephritis induced by in situ formation of immune complexes in glomerular capillary wall. Kidney int. *16:* 148–157 (1979).
10 Winkelhake, J.L.; Nicolson, G.L.: Aglycosylantibody. Effects of exoglycosidase treatments on autochthonous antibody survival time in the circulation. J. biol. Chem. *251:* 1074–1080 (1976).
11 Kozutsumi, Y.; Kawasaki, T.; Yamashina, I.: Isolation and characterization of a mannan-binding protein from rabbit serum. Biochem. biophys. Res. Commun. *95:* 658–664 (1980).
12 Köttgen, E.; Olk, B.; Kluge, F.; Gerok, W.: Gluten, a lectin with oligomannosyl specificity and the causative agent of gluten-sensitive enteropathy. Biochem. biophys. Res. Commun. *109:* 168–173 (1982).
13 Davin, J.C.; Dechenne, C.; Foidart, J.B.; Mahieu, P.: Experimental IgA nephropathy induced by IgA-lectin complexes. Xth Int. Congr. of Nephrology, London 1987, book of abstracts, p. 321.

Jean-Claude Davin, MD, Service de Pédiatrie, CHU Sart-Tilman par Liège 1, B–4000 Liège (Belgium)

Contr. Nephrol., vol. 67, pp. 117–120 (Karger, Basel 1988)

IgA Rheumatoid Factor and Other Autoantibodies in Acute Henoch-Schönlein Purpura

J.F. Knight[a], *T. Harada*[b], *M.A.B. Thomas*[b], *G. Frampton*[b], *C.O.S. Savage*[c], *C. Chantler*[a], *D.G. Williams*[b]

[a]Department of Paediatric Nephrology, Evelina Children's Hospital, Guy's Hospital; [b]Renal Unit, United Medical Schools of Guy's and St Thomas' Hospitals, Guy's Hospital; [c]Renal Unit, Royal Postgraduate Medical School, Hammersmith Hospital, London, UK

Henoch-Schönlein purpura (HSP), the most common cause of vasculitis in childhood, characteristically develops 7–10 days after an upper respiratory tract infection. Immunohistology reveals widespread deposition of IgA in the mesangium and in the small vessels of the skin and the gut [1]. Patients with acute HSP develop increased serum IgA levels [2] which are probably due to an increase in polymeric or aggregated IgA in the serum [3] (often termed IgA immune complexes, although evidence for the presence of antigen is lacking).

These observations have given rise to the view that HSP is a disease in which showers of circulating IgA immune complexes, provoked by a mucosal infection, are deposited in the basement membranes of small vessels in the affected organs, provoking monocytic infiltration and leucocytoclastic vasculitis.

The recent identification of antibodies to neutrophil cytoplasmic proteins in microscopic polyarteritis and Wegener's granulomatosis [4] has aroused interest in the role of specific autoantibodies in the pathogenesis of vasculitic diseases. An association between anti-DNA antibodies and systemic lupus erythematosus (SLE) has long been recognised. More recently, antibodies against negatively charged phospholipids such as cardiolipin have been implicated in the thrombotic manifestations of SLE [5]. Antibodies to the constant region of IgG (rheumatoid factors), a hallmark of rheumatoid arthritis, are often also found not only in SLE but also in other chronic inflammatory conditions.

It is possible that in HSP IgA antibodies are raised by antigenic determinants on the infecting organism which are cross-reactive with self antigens. These considerations have prompted us to examine plasma from cases of acute HSP for antibodies to single- and double-stranded DNA, to cardiolipin, to neutrophil cytoplasmic antigens, and to IgG.

Methods

Patients

Twenty millilitres of heparinised blood was collected from 36 children aged 15 months to 15 years (mean 7.8 years) with acute HSP. All had the characteristic rash at the time the blood was taken. Eighteen of 36 (50%) had haematuria; 10 of 36 (30%) had sufficiently severe disease to warrant renal biopsy, which showed deposition of IgA in each case. Convalescent samples were collected from 18 of these subjects.

Controls

Plasma was collected from 30 children aged 3–15 years (mean 9.3 years) attending the renal clinic. Their diagnoses included hypertension, unilateral cystic dysplasia, obstructive uropathy, adult polycystic disease and reflux nephropathy. None had a history of HSP or IgA nephropathy. Informed consent was obtained from parents and from older children. The study was approved by the Ethics Committee of the United Medical and Dental Schools of Guy's and St Thomas' Hospitals.

Specific Antibody Assays

Plasma was separated within 3 h of collection and stored at $-70\,^{\circ}C$ until tested. Anti-DNA antibodies ($n = 25$), anti-cardiolipin antibodies ($n = 13$) and anti-neutrophil cytoplasmic antibodies ($n = 20$) were measured as described [4–6]. IgA-RF was measured by the method of Sinico et al. [7].

Statistics

Differences were analysed by the Mann-Whitney U test for non-parametric data.

Results

Raised serum concentrations of IgG, IgM or IgA antibodies to double- or single-stranded DNA or to neutrophil cytoplasmic antigens were not detected in any of the patients. One child had raised IgA anti-cardiolipin antibodies; none had IgG or IgM anti-cardiolipin antibodies.

There was a significant difference in the plasma concentrations of IgA-RF between acute patient samples and controls ($p = 0.01$, fig. 1). Twelve of 36 children (33%) had IgA-RF levels above the normal range. IgA-RF values tended to return to normal in convalescent plasma (fig. 1).

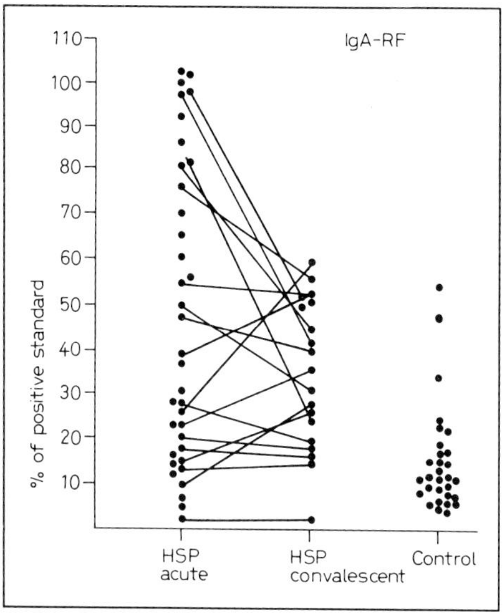

Fig. 1. IgA-RF values in HSP and control plasma.

Discussion

We have demonstrated raised IgA-RF activity in 33% of children with acute HSP. This study confirms the work of Saulsbury [8], who found IgA antibodies to IgG in 13 of 24 children with acute HSP using a gel diffusion solid-phase assay, and contrasts with the reports of Lesavre et al. [9] and Jones et al. [10], who were unable to demonstrate IgA-RF in 20 and 16 cases of HSP, respectively. Our observation that IgA-RF activity disappears as the acute disease resolves may explain these different findings. IgA-RF has also been reported in rheumatoid arthritis, in SLE and in IgA nephropathy [7]. Its presence is therefore not peculiar to HSP – it occurs in a number of immune-complex-associated disorders.

It has been suggested that RF is generated as part of the normal immune response each time antigenic stimulation is encountered. Anti-Fc-γ antibody may act in a feedback loop which regulates production of IgG – whether as a control loop or as an amplifier is not clear. As IgA is a poor activator of complement, the binding of IgA to the IgG-Fc region may reduce the pathogenicity of IgG auto-antibodies. Such IgA-IgG heteropolymers would be detected as immune complexes by most of the assays in common use.

We have been unable to demonstrate in HSP any of the three auto-antibodies which characterise other forms of vasculitis. These autoantibodies are therefore unlikely to be a secondary product of vasculitis itself. HSP may be mediated by auto-antibodies as yet unrecognised, or may not be an auto-antibody-mediated disease.

Acknowledgements

This work was supported by The Canteen Committee of The Children's Hospital, Camperdown, NSW, Australia; by the Special Trustees of Guy's Hospital, London, UK; by the National Kidney Research Fund, UK, and by the UK Medical Research Council.

References

1 Stevenson, J.A.; Leong, L.A.; Cohen, A.H.; Border, W.A.: Henoch-Schönlein purpura. Simultaneous demonstration of IgA deposits in involved skin, intestine and kidney. Archs Path. Lab. Med. *106:* 192–195 (1982).

2 Tanaka, M.; Fujisawa, S.; Okuda, R.: Anaphylactoid purpura in childhood. III. Follow-up study of serum levels of immunoglobulins (γG, γM and γA). Annls Paediat. Jap. *15:* 19–26 (1969).

3 Levinsky, R.J.; Barratt, T.M.: IgA immune complexes in Henoch-Schönlein purpura. Lancet *ii:* 1100–1103 (1979).

4 Savage, C.O.S.; Winearls, C.G.; Jones, S.; Marshall, P.D.; Lockwood, C.M.: Prospective study of radioimmunoassay for antibodies against neutrophil cytoplasm in diagnosis of systemic vasculitis. Lancet *i:* 1389–1393 (1987).

5 Frampton, G.; Hartley, B.; Cameron, J.S.: Glomerular thrombi in lupus nephritis: an association with the presence of IgG anti-cardiolipin antibodies (Abstract). Xth Int. Congr. of Nephrology, London 1987.

6 Frampton, G.; Perl, S.; Bennett, A.; Cameron, J.S.: Platelet-associated DNA and anti-DNA antibody in systemic lupus erythematosis with nephritis. Clin. exp. Immunol. *63:* 621–628 (1986).

7 Sinico, R.A.; Fornasieri, A.; Oreni, N.; Benuzzi, S.; D'Amico, G.: Polymeric IgA rheumatoid factor in idiopathic IgA mesangial nephropathy (Berger's disease). J. Immun. *137:* 536–541 (1986).

8 Saulsbury, F.T.: IgA rheumatoid factor in Henoch-Schönlein purpura. J. Paediat. *108:* 71–76 (1986).

9 Lesavre, P.; Monteiro, R.C.; Chevailler, A.; Berger, J.; Halbwachs-Mecarelli, L.; Noël, L.H.: Aspects immunologiques récents des glomérulonéphrites humaines. Actual. néphrol. *1987:* 58–73.

10 Jones, C.L.; Powell, H.R.; Kincaid-Smith, P.; Robertson, D.M.: IgA rheumatoid factor in IgA nephropathy and Henoch-Schönlein nephritis. Abstr. Australasian Soc. of Nephrology, 1987, pp. 16–17.

D. Gwyn Williams, MD, Clinical Science Laboratories, 17th Floor, Guy's Tower, Guy's Hospital, London SE1 9RT (UK)

Contr. Nephrol., vol. 67, pp. 121–124 (Karger, Basel 1988)

Unconjugated Hyperbilirubinemia due to Ciclosporin Administration in Children with Nephrotic Syndrome

A. Edefonti[a], *L. Ghio*[a], *A. Bettinelli*[a], *G. Paterlini*[a], *M. Giani*[a], *G. Nebbia*[a], *A. Aniasi*[b], *D. Cattarelli*[a]

[a]Clinica Pediatrica G. e D. De Marchi, Università di Milano;
[b]Centro Trasfusionale e di Immunologia dei Trapianti, Milano, Italia

Hyperbilirubinemia is a well-known adverse side effect of ciclosporin (CS) treatment [2]. It is generally conjugated, related to hepatic toxicity, associated with altered liver function tests and, in most reports, dose dependent [3]. Unconjugated hyperbilirubinemia and severe jaundice following CS treatment in renal transplantation is an uncommon finding and appears to be secondary to ABO autoimmune hemolytic anemia [1]. We report unconjugated hyperbilirubinemia and mild jaundice recently observed in 2 children with frequently relapsing nephrotic syndrome (NS) treated with CS. Possible pathogenetic factors are also discussed.

Patients and Methods

These cases were observed in a multicenter randomized prospective clinical study comparing 2 different treatment schedules in patients with steroid-dependent or frequently relapsing NS:

Treatment A: CS at an initial dose of 6 mg/kg/day, subsequently adjusted to maintain trough blood levels at 250–600 ng/ml, for 9 months, then tapered for 3 months.

Treatment B: Cyclophosphamide 2.5 mg/kg/day for 8 weeks. Prednisone was administered to both treatment groups to induce remission and then tapered as follows: 40, 30, 20 and 10 mg/m^2/day, each for 1 week, then 5 mg/m^2/day for 4 weeks.

Between October 1986 and April 1987, 13 children from our center entered the study after a detailed history was taken and laboratory tests performed to exclude hepatic, hematologic, neurologic and other renal disease.

None of them took medications other than the above-mentioned immunosuppressive drugs. Nine were assigned to CS, 4 to cyclophosphamide. Two of the 9 patients treated with CS developed unconjugated hyperbilirubinemia; both were 12 years old and had frequently relapsing NS of 37 and 124 months' duration, respectively.

Results and Discussion

The pattern of unconjugated hyperbilirubinemia in the 2 patients is shown in figure 1. An increase of total bilirubin, mainly due to an increase of unconjugated bilirubin, occurred within 1 month of the start of CS treatment and was higher in case 1. Bilirubin levels varied during follow-up, but hyperbilirubinemia persisted in almost all determinations. CS trough blood levels were maintained within or very near the standard therapeutic range.

The pathogenesis of unconjugated hyperbilirubinemia was investigated in both cases. Hemoglobin and reticulocyte values, peripheral blood smear, haptoglobin and direct Coomb's test were always normal, as were serum transaminases, gamma-glutamyl transpeptidase, alkaline phosphatase and cholic acid conjugate values, so that both hemolysis and hepatotoxicity could be excluded.

To complete the tests for unconjugated hyperbilirubinemia, we performed the calorie deprivation test in our 2 patients. After 2 days of reduced caloric intake, case 1 showed an increase of total bilirubin of more than 1.5 mg/dl, mainly due to unconjugated bilirubin. Case 2, on the other hand, failed to demonstrate such a marked bilirubin increase. In both cases hemolysis was again excluded by the test. The response of case 1 was that generally observed in Gilbert's syndrome, but such a condition cannot be diagnosed definitely until another calorie deprivation test can be performed after CS withdrawal. Nevertheless, apart from a possible pre-existing Gilbert's syndrome in case 1, which would be important only in determining higher bilirubin levels than in case 2, the bilirubin increase from normal to high values after CS administration, together with the absence of hemolysis or hepatotoxicity, seems to indicate that CS had a role in the pathogenesis of unconjugated hyperbilirubinemia in both patients.

Because of the very small number of subjects presenting this side effect, statistical analysis to clarify the influence of CS dosage or trough blood concentration on bilirubin levels cannot be performed. The CS total doses received in 90 days by cases 1 and 2 were, however, similar to those received by the other 7 cases without unconjugated hyperbilirubinemia

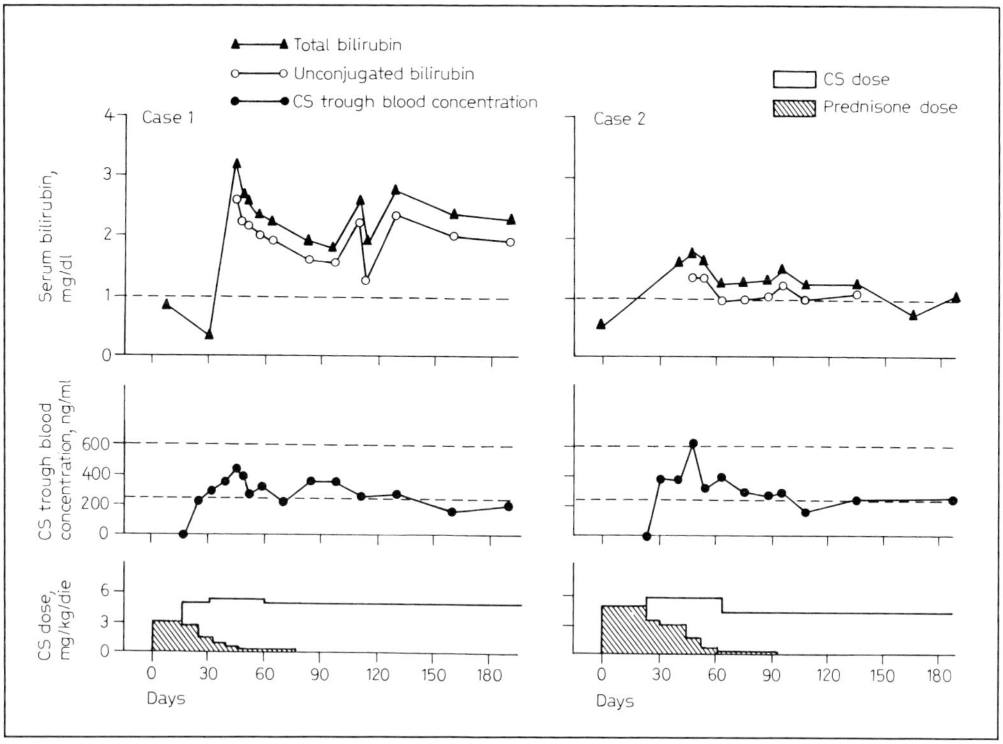

Fig. 1. Pattern of serum bilirubin, CS through blood concentration and CS dose during 6-month follow-up.

(case 1: 423.5 mg/kg; case 2: 371.4 mg/kg; other 7 cases: 463.8 mg/kg; p = n.s.). The same was true for CS trough blood levels (case 1: 318.75 ± 67.49 ng/ml; case 2: 347.5 ± 126.34 ng/ml; other 7 cases: 255.4 ± 116.5 ng/ml; p = n.s.).

Conclusion

Our observation, even though limited to 2 cases, would suggest another rare cause of hyperbilirubinemia in patients undergoing CS treatment; that is, CS interference with the hepatic uptake or conjugation of bilirubin resulting in unconjugated hyperbilirubinemia.

This up to now scantily considered side effect of CS is in addition to its well-known interference with the hepatic excretion of bilirubin, which is the main cause of the frequently reported conjugated hyperbilirubinemia of transplanted patients.

References

1 De Vecchi, A.; Zanella, A.; Egidi, F.; Ponticelli, C.: Autoimmune hemolytic anemia in a cadaveric renal transplant recipient treated with cyclosporine. Acta haemat. *73:* 216–218 (1985).

2 Rodger, R.S.C.; Turney, J.H.; Haines, I.; Michael, J.; Adu, D.; McMaster, P.: Cyclosporine and liver function in renal allograft recipients. Transplant. proc. *XV:* suppl. 1 (1983).

3 Shade, R.R.; Guglielmini, A.; Van Thiel, D.H.; Thompson, M.E.; Warty, V.; Griffith, B.; Sanghvi, A.; Bahuson, H.; Hardesty, R.: Cholestasis in heart transplant recipients treated with cyclosporin. Transplant. Proc. *XV:* suppl. 1 (1983).

A. Edefonti, MD, Nefrologia e Dialisi, Clinica Pediatrica II,
Università di Milano, Via Commenda 9, I–20122 Milano (Italy)

Neonatal Nephrology

Contr. Nephrol., vol. 67, pp. 125–129 (Karger, Basel 1988)

Management and Outcome of Renal Failure in Neonates

M. Brandis

Department of Pediatrics, Phillips-University, Marburg/Lahn, FRG

The frequency of acute renal failure in newborns is particularly high in prematures and should be one of the major concerns in pediatric nephrology. Very few studies have been done [2, 3, 8, 16, 17], and there are no valuable data on the long-term outcome of premature and neonatal renal insufficiency in later life. The incidence of late hypertension has been a subject of investigation [3] without clear information.

Recognition of disturbed renal function has to be evaluated by comparison with the normal physiological development. Fluid restriction in premature neonates has been used as a prophylactic measure against the appearance of patent ductus arteriosus [11]. In this situation, pharmacological inhibition of the prostaglandin system by indomethacin can lead to decreased renal blood flow and renal failure [11]. If it is therefore mandatory to use prostaglandin inhibitor in neonates, it should only be given if they are adequately hydrated [11].

In cases of extended delivery and long-standing hypoxia, the probability of shock-induced acute renal failure is likely [10, 14]. After fluid monitoring (balance), the possibility of nephrotoxic side effects of drugs is even more likely. Aminoglycosides, which are frequently necessary for treating septic neonates, can worsen renal function under these conditions [15]. Already under physiological conditions the tubular toxic effects can be detected by significantly increased excretion of enzymes [4].

In the newborn with extended oliguria or anuria, the temptation to use furosemide or any other diuretic might very well be justified. The use of furosemide, particularly in prematures, is, as we know, complicated by the stimulatory action of prostaglandin secretion, which might counteract the effect of indomethacin [11]. The use of furosemide should therefore be restricted in general to the volume-overloaded child at low dosages. On the

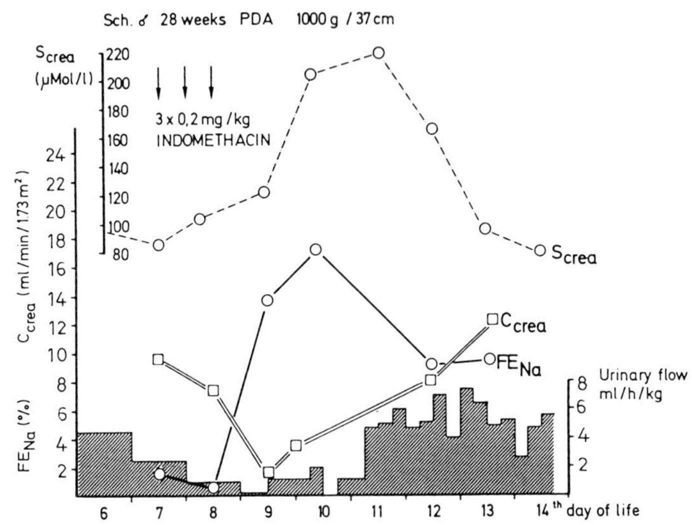

Fig. 1. Infant with acute renal failure. Clinical development.

other hand, furosemide could eventually be understood as a kidney protector. Zimmerhackl et al. [19] investigated infants with renal insufficiency with and without furosemide. They found that infants and children in renal insufficiency without furosemide excrete a high amount of lactic acid comparable to sodium diuresis. This can be interpreted as a parameter of oxygen deficiency in the area of the ascending limb of Henle's loop. In infants undergoing furosemide treatment, sodium diuresis was not paralleled by lactic acid excretion. Interpretation of these data could lead to the assumption that furosemide as a diuretic might act as a protector of the renal medulla in respect to energy consumption, and in this case diminish the amount of anaerobic metabolism. This is so far quite speculative but gives another light to the use of furosemide in the case of renal insufficiency.

Diagnostic control parameters for monitoring infants with renal failure have been well established [4, 6–9, 12]. Calculation of the term fractional excretion of sodium, values lower or higher than 2.5%, has been used as a denominator for the definition of prerenal or intrarenal failure. This is valid as long as no diuretic drug is interfering with the sodium

Table I. Glomerular filtration rate (C_{cr}) after neonatal renal failure

	Outcome	
	n	%
Total	50	100
Dead	24	48
Alive	26	52
Control (0–7 years)	12	
GFR ↓↓ (22–70 ml/min × 1.73 m^2)	7	

reabsorption and has turned out to be tremendously reliable in the follow-up of renal function. In one example (fig. 1), it is shown that a premature infant developed renal failure under indomethacin. This was paralleled by a decrease in the fractional excretion of sodium. Later on, intrarenal damage developed, and FeNa increased tremendously. The advantage of these diagnostic tools is that standard laboratory techniques, like serum creatinine, urine sodium and creatinine, and spontaneous urine specimens can be used to evaluate this important quality of renal function. More sophisticated renal tubular function parameters like microprotein excretion, enzyme excretion or, as mentioned, lactate excretion, are of scientific interest in respect to their mechanisms but are not necessary for practical clinical use [4].

Some of the most important questions are first the actual outcome of acute renal failure in neonates, and second the long-term prognosis. The first question has been studied quite extensively in the last 10 years [1–6, 10, 13, 14, 18]. In our own survey in the last 7 years we diagnosed 50 children (table I) with the diagnosis of acute renal failure. Twenty-four died during the neonatal period and 26 survived. From these, 12 could later be investigated from 9 months to 6.5 years of age. A decreased creatinine clearance was measured in 7 of these 12 children. In 3 of these 7 repeated investigation revealed the same results (table I). These are very preliminary data and have been derived after very difficult follow-up studies, since some of these children have other sequelae from the neonatal intensive care treatment. The investigation is so far quite preliminary, and, surprisingly enough, only very few reports of long-term investigations exist [1, 8, 16, 17]. Nevertheless, it seems to be important to follow this question in the future.

Morphological reinvestigation showed no distinct pathological findings in ultrasonography, but the time after the incidence of acute renal failure in these children may have been too short for a long-term evaluation of kidney growth.

In summary, the management of acute renal failure in neonates consists of fluid regulation, drug monitoring, and circulatory support. The outcome of acute renal failure in neonates depends on various other organic manifestations in these children. For the survivors, the development of renal function may, in addition to other sequelae of organ damage, be one of the serious questions to be followed. It is recommended to control these children quite closely in later childhood with respect to their renal function.

References

1 Anand, S.K.; Northway, I.D.; Cruski, F.J.: Acute renal failure in newborn infants. J. Pediat. *92:* 985–988 (1978).

2 Anand, S.K.: Acute renal failure in the neonates. Pediat. Clins N. Am. *29:* 791–800 (1982).

3 Burghard, R.; Leititis, J.U.; Brandis, M.: Nierenfunktionsstörungen und akutes Nierenversagen bei Neugeborenen. Mschr. Kinderheilk. *135:* 10–21 (1987).

4 Burghard, R.; Leititis, J.U.; Gordjani, N.; Brandis, M.: Renal tubular function in prematures: Postnatal changes and influence of tubulotoxic drugs. Int. J. pediat. Nephrol. *5:* 113 (1984).

5 Cochat, P.; Bourgeois, J.; Gilly, J.; Cottin, X.; Larbre, F.; Bethenod, M.: Etude anatomique des reins de nouveau-nés décédés au décours d'un état septique. Pédiatrie *41:* 7–15 (1986).

6 Ellis, D.; Gartner, I.C.; Galvis, A.G.: Acute renal failure in infants and children: diagnosis complication and treatment. Crit. Care Med. *9:* 607–617 (1981).

7 Ellis, E.N.; Arnold, W.C.: Use of urinary indexes in renal failure in the newborn. Am. J. Dis. Child. *136:* 615–617 (1982).

8 Engle, W.D.: Evaluation of renal function and acute renal failure in the neonate. Pediat. Clins N. Am. *33:* 129–151 (1986).

9 Grylack, L.; Medani, C.; Hultzen, C.; Sivasubramanian, K.: Nonoliguric acute renal failure in the newborn. A prospective evaluation of diagnostic indexes. Am. J. Dis. Child. *136:* 518–520 (1982).

10 Kojima, T.; Kobyasahi, T.; Matsuzaki, S.; Iwase, S.; Kobayashi, Y.: Effects of perinatal asphyxia and myoglobinuria on development of acute, neonatal renal failure. Archs Dis. Childh. *60:* 908–912 (1985).

11 Leititis, J.U.; Burghard, R.; Gordjani, N.; Wildberg, A.; Seyberth, H.W.; Brandis, M.: Effect of a modified fluid therapy on renal function during indomethacin treatment for persistent ductus arteriosus Botalli. Acta paediat. scand. (in press, 1987).

12 Mathew, O.P.; Jones, A.S.; James, E.; Bland, H.; Groshong, T.: Neonatal renal failure: Usefulness of diagnostic indices. Pediatrics, Springfield *65:* 57–60 (1980).
13 McCormic, A.; Stark, M.B.: Acute renal failure of the neonate. Dimens. crit. Care Nurs. *5:* 155–161 (1986).
14 Rahman, N.; Boineau, F.G.; Lewy, J.E.: Renal failure in the perinatal period. Clin. Perinatol. *8:* 241–250 (1981).
15 Sirinavin, S.; McCracken, G.H., Jr.; Nelson, J.D.: Determining gentamicin dosage in infants and children with renal failure. J. Pediat. *96:* 331–334 (1980).
16 Sapleton, F.B.; Jones, D.P.; Green, R.S.: Acute renal failure in neonates. Incidence, etiology and outcome. Pediat. Nephrol. *1:* 314 (1987).
17 Stark, H.; Geiger, R.: Renal tubular dysfunction following vascular accidents of the kidneys in the newborn period. J. Pediat. *83:* 933–940 (1973).
18 Trompeter, R.S.; Chantler, C.; Haycock, G.B.: Tolazoline and acute renal failure in the newborn. Lancet *i:* 1219 (1981).
19 Zimmerhackl, L.B.; Kramer, M.; Blum, L.; Küster, P.; Rausch, P.; Steidel, U.; Brandis, M.: Mikrozirkulation des Nierenmarkes: Laktat-Handlung als Parameter des medullären Zellstoffwechsels bei renalen Funktionsstörungen (in press).

Prof. M. Brandis, Department of Pediatrics, Phillips University,
Deutschhausstrasse 12, D-3550 Marburg/Lahn (FRG)

Contr. Nephrol., vol. 67, pp. 130–137 (Karger, Basel 1988)

Distal Tubular Sodium Handling in Human Neonates: Clearance Studies

B.S. Arant, Jr.

Division of Pediatric Nephrology, University of Texas Health Science Center, Dallas, Tex., USA

Nearly 40 years ago, Dean and McCance [1] concluded from studies which in no way could be considered physiologic that human infants could not excrete an intravenous sodium chloride load as efficiently as adults. It is of interest to note that the dose of 10% NaCl solution given intravenously over a brief interval was similar to that given to rabbits by Sotos et al. [2] and cats by Kravath et al. [3] to investigate the role of *acute* changes in plasma osmolarity in hypertonic brain injury. Ever since, investigators have attempted to elucidate developmental mechanisms to explain this difference; moreover, interpretations of results have often been biased towards finding such a difference [4]. Although the observation has not yet been disproven, it is hyponatremia, not hypernatremia, and renal sodium wasting, not sodium retention, which confound the clinician treating preterm infants most often.

Current understanding of renal sodium handling in human neonates can be attributed to three teams of investigators: Aperia et al. [5, 6], Sulyok et al. [7] and Rodriguez-Soriano et al. [8] among whom there is general agreement. A summary of their conclusions from clearance studies is that distal tubular sodium reabsorption, measured as free water clearance (C_{H_2O}) factored for glomerular filtration rate (GFR), parallels changes in distal sodium delivery, measured as C_{H_2O} plus the clearance of sodium (C_{Na}) factored for GFR. Moreover, preterm infants appear to reject a greater fraction of filtered sodium from the proximal tubule than full-term infants, which results in more filtered sodium reaching the distal nephron. When the capacity of distal tubular mechanisms for sodium reabsorption is exceeded, a reduced fraction of sodium delivered there is reabsorbed, and fractional excretion of sodium (FE_{Na}) increases, which is observed more commonly in preterm infants. However, clearance studies used to measure distal sodium handling in human neonates have been criticized variously for failing to adhere to strict criteria required for calculation of C_{H_2O} = urine flow rate (V) – osmolar clearance (C_{osm}) to be considered valid.

What Are These Criteria?

According to Danovitch [9], clearance studies must be performed during maximal water diuresis – urinary osmolarity (U_{osm}) should be < 60 mosm/l and the fractional volume of glomerular filtrate reaching the collecting duct (V/GFR) should be $\geq 5\%$. Moreover, fractional delivery of sodium to the distal nephron should be at least 8% of the filtered load – adequate to test the capacity of distal reabsorptive mechanisms. Calculation of C_{H_2O} should represent only sodium chloride reabsorption which occurs beyond the proximal straight tubule. The accuracy of this calculation can be influenced by several factors. The resulting value can be artifactually lowered if V is reduced by vasopressin-mediated water reabsorption from the collecting duct. If sodium were reabsorbed in the distal tubule in exchange for potassium or hydrogen ion, the C_{Na} will be decreased and the calculation of distal sodium delivery (C_{H_2O} + C_{Na}) artifactually reduced. For this reason some investigators have used C_{Na} plus the clearance of potassium to calculate distal sodium delivery. When bicarbonate is reabsorbed in the distal nephron, hydrogen ion will have been secreted and sodium reabsorbed – another error in calculations because C_{Na} will have decreased. The clearance of chloride (C_{Cl}) is the only term not subject to errors for calculating distal tubular sodium handling during water diuresis by clearance techniques. Finally, extracellular fluid volume (ECFV) expansion lowers distal tubular sodium reabsorption, even when every other criterion for clearance methodology has been met. Danovitch and Bricker [10] demonstrated this effect most clearly in conscious dogs, as Martinez-Maldonado et al. [11] had done previously in anesthetized rats, whereby progressive saline expansion resulted in less of the sodium delivered distally to be reabsorbed. Euvolemia, therefore, is required for valid calculations. The newborn infant poses a major problem in this regard because of differences in ECFV among infants of different conceptional ages (CA) at birth [12] as well as between infants of the same CA studied at different postnatal ages when ECFV would have changed from birth depending upon the timing and extent of spontaneous diuresis and fluid balance [13, 14].

Did Studies Reported Previously in Human Infants Conform to Criteria Mandated for Valid Clearance Methodology?

Aperia et al. [6] studied 5 preterm infants during the first week of life; U_{osm} was not stated, but V/GFR was > 5% and the distal delivery of sodium was > 8%. Five full-term infants studied during the first 2 weeks of life, 1 in the first week and 4 in the second week, had mean U_{osm} of 55 mosm/l, V/GFR > 5% and distal deliveries of sodium > 8% [5]; the 'sodium term' was used for both studies. Sulyok et al. [7] studied 8 preterm and 10 full-term infants at 6 days of age. Mean U_{osm} was 161 and 144 mosm/l, respectively, V/GFR was just > 5% in preterm infants but < 5% in full-term infants and distal delivery of sodium was < 8% in both groups; the 'sodium term' was used for calculations. Rodriguez-Soriano et al. [8] studied 13 preterm and 15 full-term infants at 7 days of age whose mean U_{osm} were 67 and 52 mosm/l, respectively, V/GFR was > 5%, and distal sodium delivery was > 8%; 'sodium', 'sodium plus potassium', and 'chloride terms' were used.

Does Failure to Achieve Maximum Water Diuresis, as Was the Case in Some Previous Reports, Invalidate Conclusions Made from Clearance Studies of Human Infants?

To address this issue, we studied infants exhibiting spontaneous diuresis during the first week of life whose plasma sodium concentrations (130–150 mEq/l) and osmolarities (270–295 mosm/l) were normal, whose U_{osm} were ≤250 mosm/l and whose V/GFR were > 5%. In addition, no infant was included who had been treated since birth by volume expansion, whose weight at study was greater than birth weight, or who had received a diuretic. Forty-nine infants met these criteria. Birth weights were between 780 and 3,880 g, gestational ages between 29 and 40 weeks, and postnatal ages between 1 and 7 days. Timed urine collections were made, and plasma and urinary sodium, chloride, creatinine and osmolarity were measured by methods described previously [15]. These infants were divided arbitrarily

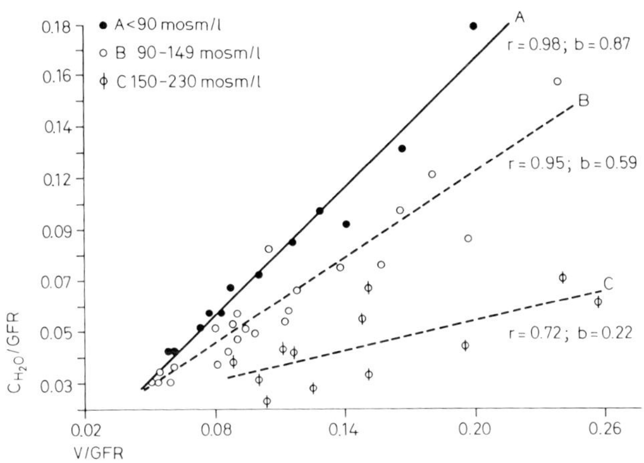

Fig. 1. The influence of urinary osmolarity on calculations of sodium handling in the distal nephron by clearance methodology is compared. Distal tubular sodium reabsorption (C_{H_2O}/GFR) is compared with the 'volume term' for distal sodium delivery (V/GFR). The correlation coefficient (r) and slope (b) for each regression line are stated for each group.

into three groups: *group A:* 12 infants had U_{osm} osmolarities between 30 and 83 mosm/l – these infants were considered similar to the full-term infants studied by Aperia et al. [5] and to both the preterm and full-term infants studied by Rodriguez-Soriano et al. [8]; *group B:* 25 infants with U_{osm} between 92 and 148 mosm/l; *group C:* 12 infants with U_{osm} between 150 and 237 mosm/l – the infants in these latter two groups were similar to those studied by Sulyok et al. [7].

Distal sodium reabsorption (fig. 1) varied directly with V/GFR or 'volume term' in each of the three groups. It was readily apparent; however, in group A (infants undergoing maximum water diuresis) the regression line had the steepest slope (0.87) which was similar to the line of identity. Although the correlation remained statistically significant for groups B (r = 0.95) and C (r = 0.72), the higher the U_{osm}, the lower the slope of the regression line (B = 0.59; C = 0.22). When the 'sodium term' was used, similar relationships were noted. Finally, when the 'chloride term' was used, both correlation coefficients (A = 0.97; B = 0.96) and the slopes of the lines (A = 0.62; B = 0.63) were similar for groups A and B; however, the correlation coefficient in group C was remarkably lower ($r = 0.75$) and the slope of the line ($b = 0.32$) only half that of groups A and B. The higher slopes with volume and sodium terms compared to the chloride term may have been due to methodologic differences in chloride measurements, to inaccuracies in measurement of chloride, to sodium exchanged for potassium and hydrogen ion in the distal nephron or to vasopressin effect. Correlation coefficients in the intermediate U_{osm} groups for each term of sodium delivery were similar, as are the slopes of the regression lines. The group with the highest U_{osm} had comparably lower correlation coefficients, as did the lowest slopes to the lines to describe the relationship to C_{H_2O}. It appeared, therefore, that $U_{osm} < 150$ mosm/l would permit the use of clearance methodology to investigate distal tubular sodium handling when the 'chloride term' is used.

Does Distal Tubular Sodium Reabsorption Vary with Conceptional Age?

Distal tubular sodium handling using the 'chloride term' was compared with CA for the 37 infants with $U_{osm} < 150$ mosm/l. Distal sodium reabsorption (C_{H_2O}/GFR) was highest in the least-mature infants and decreased with CA ($r = -0.46$, $p < 0.01$). Distal sodium delivery ($C_{Cl} + C_{H_2O}$/GFR) also decreased with CA ($r = -0.53$, $p = 0.001$). The fractional reabsorption of sodium delivered to the distal nephron ($C_{H_2O}/C_{Cl} + C_{H_2O}$) increased with conceptional age ($r = 0.52$, $p = 0.001$). Finally, V/GFR was also observed to decrease with CA ($r = -0.61$, $p < 0.001$), i.e. preterm infants excreted a greater fractional volume of glomerular filtrate in their urines than did full-term infants.

Furthermore, the infants were divided arbitrarily into those <35 weeks CA and those >35 weeks CA – the same age groups used by Rodriguez-Soriano et al. [8]. By this comparison, distal sodium reabsorption (C_{H_2O}/GFR) increased pari passu with distal sodium delivery ($C_{Cl} + C_{H_2O}$/GFR) regardless of CA ($r = 0.97$, $p < 0.001$). The fraction of sodium delivered to the distal nephron that was reabsorbed ($C_{H_2O}/C_{Cl} + C_{H_2O}$) was compared with distal sodium delivery ($C_{Cl} + C_{H_2O}$/GFR) (fig. 2). The negative correlation ($r = -0.37$) was less strong but remained statistically significant ($p = 0.025$). Distal tubular sodium reabsorption in infants <35 weeks CA appeared similar to those >35 weeks CA. Distal sodium deliv-

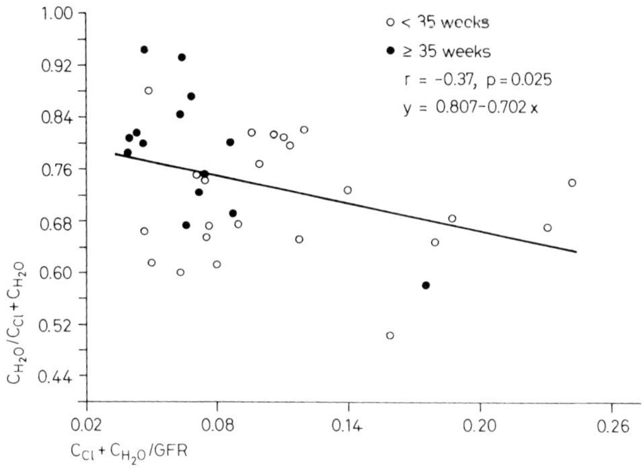

Fig. 2. The fractional reabsorption of sodium delivered to the distal nephron is compared with the distal delivery of sodium calculated by clearance methodology as the 'chloride term'.

ery varied directly with V/GFR in a manner that was similar for infants <35 weeks CA ($r = 0.90$) and those >35 weeks CA ($r = 0.91$); the slope of the line was 1.0 when all 37 infants were considered together.

Which, Then, Is the More Important Factor in Distal Tubular Sodium Handling in Newborn Infants, Conceptional Age or Fractional Urine Flow Rate?

When the influence of CA or V/GFR was isolated for statistical analysis by the method of partial correlation [16], the statistical significance of the correlation between CA and distal tubular sodium reabsorption (C_{H_2O}/GFR) or sodium delivery ($C_{Cl} + C_{H_2O}$/GFR) was lost; however, the strength of correlation between V/GFR and either measure of distal sodium handling was unchanged.

Conclusions

Distal tubular sodium reabsorption, measured as C_{H_2O}/GFR during spontaneous water diuresis in euvolemic human neonates, increased pari passu with the distal delivery of sodium – regardless of the CA of the infant. Secondly, the fraction of filtered sodium chloride delivered to the distal nephron was greater in preterm than in full-term infants due to differences in fractional reabsorption of sodium chloride in the proximal nephron segments. These findings are not different from those of previous clearance studies [5–8]; however, proximal tubular sodium reabsorption and distal delivery of sodium were related directly to V/GFR which, in turn, was related inversely to CA – giving the impression of qualitative differences in tubular sodium handling during development.

Speculation

It would be reasonable to speculate, therefore, that the apparent developmental differences in the renal sodium handling in human neonates can be explained, at least in part, by differences in ECFV of neonates at the time of study, which causes an increased volume of glomerular filtrate to be excreted as urine: first, there is greater rejection of sodium and water from the proximal tubule with increased distal delivery of sodium which, at some point, overwhelms the reabsorptive capacity of the distal nephron; thereby, the fraction of sodium chloride delivered distally that is reabsorbed is reduced and FE_{Na} is increased. This same renal response to ECFV expansion has been reported for kidneys of adult rats [11], dogs [10] and humans [17]. The relatively higher ECFV, normal for preterm infants, can, if perpetuated following birth by aggressive fluid therapy, prevent the reduction in ECFV corresponding to the usual weight loss in the first days of postnatal life. Consequently, there is renal sodium wasting, negative sodium balance and a requirement for sodium chloride supplementation. On the other hand, if ECFV in preterm infants decreases appropriately following birth, more sodium chloride will be reabsorbed proximally, distal delivery reduced, the distal capacity to reabsorb sodium not exceeded, the fractional reabsorption of sodium delivered distally increased, and FE_{Na} decreased. Through this mechanism of physiologic adaptation of the fetus to postnatal life, inappropriate sodium wasting is avoided and hyponatremia prevented. Thus, the search for another mechanism to explain developmental differences in distal sodium handling is no longer clinically relevant.

References

1 Dean, R.F.A.; McCance, R.A.: The renal responses of infants and adults to the administration of hypertonic solutions of sodium chloride and urea. J. Physiol. *109:* 81–97 (1949).
2 Sotos, J.F.; Dodge, P.R.; Meara, P.; Talbot, N.B.: Studies in experimental hypertonicity. I. Pathogenesis of the clinical syndrome, biochemical abnormalities and cause of death. Pediatrics *26:* 925–938 (1960).
3 Kravath, R.E.; Aharon, A.S.; Abal, G.; Finberg, L.: Clinically significant physiologic changes from rapidly administered hypertonic solutions: acute osmol poisoning. Pediatrics, Springfield *46:* 267–275 (1970).
4 Spitzer, A.: The role of the kidney in sodium homeostasis during maturation. Kidney int. *21:* 539–545 (1982).
5 Aperia, A.; Broberger, O.; Thodenius, K.; Zetterstrom, R.: Renal response to an oral sodium load in newborn full-term infants. Acta paediat. scand. *61:* 670–676 (1972).
6 Aperia, A.; Broberger, O.; Thodenius, K.; Zetterstrom, R.: Developmental study of the renal response to an oral salt load in preterm infants. Acta paediat. scand. *63:* 517–524 (1974).
7 Sulyok, E.; Varga, F.; Gyory, E.; Csaba, I.F.: On the mechanism of renal sodium handling in newborn infants. Biol. Neonate *37:* 75–79 (1980).
8 Rodriguez-Soriano, J.; Vallo, A.; Oliveros, R.; Castillo, G.: Renal handling of sodium in premature and full-term neonates. A study using clearance methods during water diuresis. Pediat. Res. *17:* 1013–1016 (1983).
9 Danovitch, G.M.: Clearance methodology in the study of the function of the distal tubule. Renal Physiol., Basel *1:* 56–60 (1978).
10 Danovitch, G.M.; Bricker, N.S.: Influence of volume expansion of NaCl reabsorption in the diluting segments of the nephron. A study using clearance methods. Kidney int. *10:* 229–238 (1976).
11 Martinez-Maldonado, M.; Eknoyan, G.; Suki, W.N.: Influence of volume expansion on renal diluting capacity in the rat. Clin. Sci. mol. Med. *46:* 331–345 (1974).
12 Friis-Hansen, B.: Body water compartments on children. Changes during growth and related changes in body composition. Pediatrics *28:* 169–181 (1961).
13 Stonestreet, B.R.; Bell, E.F.; Warburton, D.; Oh, W.: Renal response in low-birth-weight neonates. Am. J. Dis. Child. *137:* 215–219 (1983).
14 Shaffer, S.G.; Bradt, S.K.; Meade, V.M.; Hall, R.T.: Extracellular fluid volume changes in very low birth weight infants during first 2 postnatal months. J. Pediat. *111:* 124–128 (1987).
15 Arant, B.S., Jr.: Developmental patterns of renal functional maturation compared in the human neonate. J. Pediat. *92:* 705–712 (1978).
16 Snedecor, G.W.; Cochran, W.G.: Statistical methods; 7th ed., pp. 361–363 (Iowa State University Press, Ames 1980).
17 Buckalew, V.M., Jr.; Walker, B.R.; Puschett, J.B.; Goldberg, M.: Effects of increased sodium delivery on distal tubular sodium reabsorption with and without volume expansion in man. J. clin. Invest. *49:* 2336–2344 (1970).

Billy S. Arant, Jr., MD, Department of Pediatrics, University of Texas,
Health Science Center, 5323 Harry Hines Boulevard, Dallas, TX 75235-9063 (USA)

Contr. Nephrol., vol. 67, pp. 138–144 (Karger, Basel 1988)

Role of Atrial Natriuretic Peptide in the Postnatal Adaptation of Term and Preterm Infants

T. Tulassay

First Department of Pediatrics, Semmelweis University Medical School, Budapest, Hungary

Atrial natriuretic peptide (ANP) has been shown to be ultimately involved in the regulation of extracellular fluid volume (ECFV) by increasing renal sodium and water excretion and GFR [1, 2]. ANP interacts with several vasoactive hormone systems which are known to regulate sodium and water balance [3, 4]. Increase in ECFV is associated with immediate ANP release [5], while severe volume depletion inhibits the hormone secretion [6].

Rapid changes in body fluid volume with predominant decrease in ECFV occurs physiologically in newborns during early postnatal adaptation [7, 8]. Preterm infants have higher total body fluid volume [8]. Disturbed cardiopulmonary adaptation which occurs frequently in this population, may delay the isotonic contraction of ECFV. The known association between changes in ECFV on the one hand, and plasma level of ANP on the other, raises the question whether ANP is involved in the regulatory process of ECFV also in the early postnatal adaptational period. Therefore, our investigations were focussed on this question.

Patients and Method

We measured the plasma concentrations of ANP by special radioimmunoassay [9] in the following groups.

Study 1. Twenty full-term neonates born by vaginal delivery after uncomplicated pregnancies were studied immediately after birth and on the 3rd, 5th, 7th and 10th days of life.

Study 2. ANP was measured in 18 premature infants and 38 full-term neonates during the first week of life (mean age 3 days, range 1–7, for full-term infants and 3 days, range 2–7, for premature infants). All were delivered vaginally without any complications. Apgar scores at 1 min were ≥ 7.

Study 3. Eleven premature infants (GA: 32 ± 1.2 weeks; BW: 1,500 ± 340 g) were investigated for 5 weeks after birth to demonstrate the natural course of changes in plasma ANP level in prematures.

Study 4. Sixteen 1-week-old premature infants were randomized into two groups. Infants in both groups had comparable birth weight (1,571 ± 212 vs. 1509 ± 157 g), gestational age (31.1 ± 0.8 vs. 31.1 ± 0.9 weeks) and showed no signs of cardiopulmonary adaptational disturbance. In group 1 (n = 9) infants were fed with human milk or a formula adapted to human milk, they had a sodium intake of 1.5 mmol/kg/day. In group 2 (n = 7) additional sodium supplementation of 4–5 mmol/kg/day was applied. Sodium balance and plasma concentration of ANP were measured at weekly intervals for 5 weeks.

Results and Discussion

In study 1 ANP concentrations in venous cord blood did not differ from those obtained 1 h after birth being 22.5 ± 2.5 and 22.1 ± 3.5 fmol/ml, respectively. These values are within the normal range of ANP plasma concentrations ($\bar{x}$ = 23.9 ± 1.2 fmol/ml; n = 91, age range 4 weeks to 18 years). On the third day of life there was a significant increase to 45.2 ± 5.4 fmol/ml ($p < 0.005$). ANP plasma levels began to decline on the fifth day of life, but remained significantly high (31.4 ± 3.5 fmol/ml, $p < 0.025$). On the 10th day, ANP concentrations decreased below the normal values (14.6 ± 2.1 fmol/ml, $p < 0.05$). There was a continuous decline in body weight which reached its maximum on the fifth day of life (a difference of 8% compared with the birth weight). Body weights started to rise from the 7th day when ANP values were within the normal range. Similar results have been obtained by Weil et al. [10] and Shaffer et al. [11]. Dolan [12], measuring the immunoreactive ANP content per milligram protein in the atrial tissue, found higher values during fetal life and the ANP content decreased significantly after birth. These data indicate an enhanced ANP release from atrial myocytes into the circulation shortly after birth. Therefore, we suggest that ANP may be one of the factors which mediates the postnatal changes of body fluid volume. The mechanism shown in figure 1 is proposed.

In study 2 plasma levels of ANP were compared in healthy full-term and premature infants. Premature infants had markedly elevated plasma

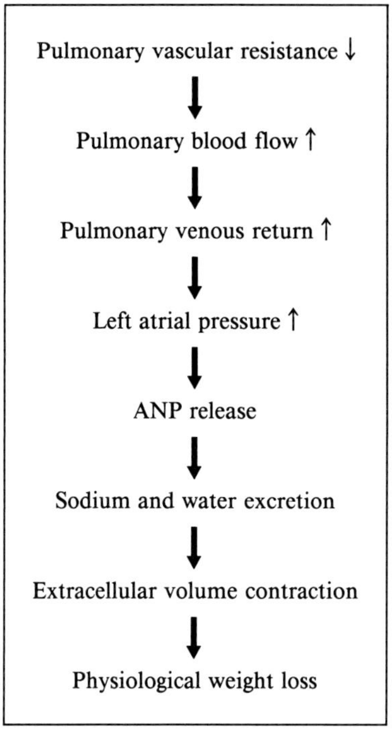

Fig. 1. Proposed pathomechanism of extracellular volume contraction after birth.

ANP levels during the first week (173.4 ± 45.2 vs. 46.7 ± 8.0 fmol/ml, $p < 0.01$). The values in full-term newborns were about twice those obtained from older children. This is compatible with the observation in study 1.

The premature infants studied here had no signs of cardiopulmonary disturbances which may result in an increased atrial pressure and a concomitant ANP excess. This was demonstrated by Shaffer et al. [11] who found higher ANP levels in the plasma in newborns with respiratory distress syndrome. Moreover, Rascher and Seyberth [13] showed a clear association between elevated plasma concentration of ANP and symptomatic patent ductus arteriosus. It is of interest, however, that although ANP values dropped after surgical closure of the ductus, they remained elevated by about sixfold. These observations also support previous findings [14] demonstrating that premature infants with a gestational age ≤32 weeks have elevated basic levels of ANP in the plasma.

The possibility still remains that this is the result of a passive accumulation of the hormone originated from the mother. Placenta has been found to be a rich source of ANP receptors and also ANP can cross it from the mother to the fetus [15]. However, Yamaji et al. [16] measured the ANP concentration in cord arterial and venous and maternal venous plasma and demonstrated that the ANP levels were higher in umbilical cord arterial than venous, or maternal venous plasma, indicating that ANP may be secreted by human fetal atrial tissue.

In an experimental work, Wey et al. [17] demonstrated that the fetal cardiac ANP system in the rat does respond to changes in intracardiac pressure. Administration of indomethacin produced a striking increase in fetal plasma ANP and a simultaneous decrement in the ANP concentration of the right atrium [17].

It has been also shown that extraatrial sources may contribute to the plasma ANP level: a ventricular ANP synthesis was demonstrated during fetal life [17]. It decreases with age while ANP synthesis in the atrial myocytes increases.

Based on these data it is evident that the elevated ANP levels measured in premature infants are the result of an actively working cardiac hormone system. One may hypothesize that ANP has a certain physiological role in hemodynamic adaptation during the postnatal period.

To understand this role more precisely, we investigated 11 premature infants without any signs of cardiopulmonary disturbances for 5 weeks after birth (study 3). Premature infants had a tenfold elevation in their plasma ANP levels 2 days after birth compared to controls (252.6 ± 63.3 vs. 23.9 ± 12 fmol/ml). Plasma ANP values decreased gradually reaching the normal range only at the third postnatal week. The decrement in plasma ANP levels was accompanied by the decrease in urinary sodium excretion during this period and a significant positive correlation was found between the two variables ($y = 7.2 \pm 11.8$, $r = 0.8$, $p < 0.001$). ANP levels correlated not only with urinary sodium excretion but also showed a negative correlation with the systolic blood pressure during the first 5 weeks of life (1st week: $r = -0.76$, $p < 0.001$; 2nd week: $r = -0.78$, $p < 0.001$; 3rd week: $r = -0.86$, $p < 0.001$; 4th week: $r = -0.97$, $p < 0.001$; 5th week: $r = -0.82$, $p < 0.001$).

It is concluded from this study that ANP has a determining role in balancing sodium homeostasis and regulates blood pressure in premature infants during the adaptational period to extrauterine life.

The 'late' hyponatremia of preterm infants is related to immaturity of the developing kidney [18]. It can be avoided by the supplementation of sodium intake [19], which also prevents secondary hormonal changes [19].

Therefore, in study 4 measurements of ANP concentrations in the plasma were carried out in two groups of infants with and without additional sodium supplementation.

Sodium intake in group 1 varied from 1.4 to 1.8 mmol/kg/day throughout the study. This low intake resulted in a relatively constant, low urinary sodium excretion and a negative sodium balance during the first 2 weeks. Serum sodium concentration fell; the lowest value of 131 ± 2.2 mmol/l was noted during the third week.

In group 2, daily sodium supplementation of 4.6 ± 1.0 mmol/kg caused a significant increase in urinary sodium excretion. A decrease in serum sodium concentration did not occur, and sodium balance became positive during the second week of the study. The plasma concentration of ANP did not differ at the beginning of the study. In the nonsupplemented group the ANP concentration decreased gradually, reaching its lowest value during the third study week, when the minimum serum sodium concentration was also observed. The mean plasma concentration of ANP in the nonsupplemented group reached the normal value for healthy, full-term, 7- to 10-day-old newborn babies (n = 14, plasma ANP 20.8 ± 2.2 fmol/ml) during the third week of life.

Plasma concentration of ANP increased by 30% after sodium supplementation in group 2 and remained practically constant. The differences between supplemented and nonsupplemented groups from the second week of life were statistically significant. ANP values in supplemented preterm infants were also higher than those of healthy, full-term newborn infants at the second through fifth weeks.

Our data for postnatal changes of plasma ANP values in premature infants correspond to the alterations of other hormone systems described previously. In children not receiving supplemental sodium, the ANP concentration in serum was high at 1 week of life, and a steady decrease was observed thereafter. The serum ANP concentration at the second week was significantly lower than values in premature infants obtained shortly after birth, and reached the normal range during the third week of life. The decrease in plasma ANP concentration was accompanied by the decrease in urinary sodium excretion.

Supplementation of sodium, on the other hand, caused a 30% elevation in plasma ANP concentration to a level significantly higher than normal values in full-term babies or older children. In the supplemented group, higher urinary sodium excretion occurred: the exact mechanism of this phenomenon is not known.

The elevated plasma ANP values induced by additional sodium intake are not able to produce an appropriate increase in natriuresis in premature infants. If our data are compared with those Sagnella et al. [20] obtained in adults ingesting low, normal, and high sodium diets, it is obvious that the premature infant kidney is less responsive to ANP. Doubling the ANP level in adults ingesting a high sodium diet resulted in an adequate balance in sodium homeostasis, whereas in premature infants elevated plasma concentrations of ANP by more than twofold caused an increase in sodium excretion, while still maintaining positive sodium balance. In the supplemented infants, mean cumulative sodium retention of 52 mmol/kg occurred during the study period. Positive sodium balance is a characteristic of sodium metabolism in the growing organism. The accretion of body tissue, particularly bone, is associated with the retention of sodium. It is possible that the reduced natriuretic effect of ANP in premature infants is a response of the developing kidney to conserve sodium efficiently for growth.

The mechanism by which higher sodium intake in premature infants is accompanied by a rising level of ANP is not known. Elevated vascular or extracellular volume, with an increase in atrial pressure, was shown to cause an immediate ANP release mediated by stretch receptors in the atria. The extracellular volume was not determined in our study, but we can speculate that increased ANP secretion would be a consequence of sodium-induced expansion of the extracellular fluid volume.

The hormonal alterations we describe appear to be secondary adaptive consequences, possibly helping to maintain sodium balance. ANP may be an important hormonal system for regulating sodium homeostasis in premature infants.

References

1 Needleman, P.; Greenwald, J.E.: Atriopeptin: a cardiac hormone intimately involved in fluid, electrolyte and blood pressure homeostasis. New Engl. J. Med. *314:* 828–834 (1986).

2 Ballermann, B.J.; Brenner, B.M.: Biologically active atrial peptides. J. clin. Invest. *76:* 2041–2048 (1985).

3 Cody, R.J.; Atlas, S.A.; Laragh, J.H.; Kubo, S.H.; Covit, A.B.; Ryman, K.S.; Shaknovich, A.; Pondolfino, K.; Clark, M.; Camargo, M.J.F.; Scarborough, R.M.; Lewicki, J.A.: Atrial natriuretic factor in normal subjects and heart failure patients. J. clin. Invest. *78:* 1362–1374 (1986).

4 Samson, W.K.: Atrial natriuretic factor inhibits dehydration and hemorrhage-induced vasopressin release. Neuroendocrinology *40:* 277–279 (1985).

5 Singer, D.J.; Shore, A.C.; Markandu, N.D.; Buckley, M.G.; Sagnella, G.A.; MacGregor, G.A.: Dissociation between plasma atrial natriuretic peptide levels and urinary sodium excretion after intravenous saline infusion in normal man. Clin. Sci. *73:* 285–289 (1987).
6 Tulassay, T.; Rascher, W.; Körner, A.; Miltényi, M.: Atrial natriuretic peptide and other vasoactive hormones during treatment of severe diabetic ketoacidosis in children. J. Pediat. *111:* 329–334 (1987).
7 Friis-Hansen, B.: Changes in body water compartments during growth. Acta paediat. scand. *110:* suppl. 46, pp. 1–67 (1957).
8 Bell, E.F.; Oh, W.: Fluid and electrolyte balance in very low birth weight infants. Clin. Perinat. *6:* 139–150 (1979).
9 Rascher, W.; Tulassay, T.; Lang, R.E.: Atrial natriuretic peptide in plasma of volume-overloaded children with chronic renal failure. Lancet *ii:* 303–305 (1985).
10 Weil, J.; Bidlingmaier, F.; Döhlemann, C.; Kuhnle, U.; Strom, T.; Lang, R.E.: Comparison of plasma atrial natriuretic peptide levels in healthy children from birth to adolescence and in children with cardiac diseases. Pediat. Res. *20:* 1328–1331 (1986).
11 Shaffer, S.G.; Geer, P.G.; Goetz, K.L.: Elevated atrial natriuretic factor in neonates with respiratory distress syndrome. J. Pediat. *109:* 1028–1033 (1986).
12 Dolan, L.M.: Alterations in right an left atrial immunoreactive atrial natriuretic polypeptide (IR-ANP) concentration (CONC) at birth. BAAP Congress, New York 1987, abstr.
13 Rascher, W.; Seyberth, H.W.: Atrial natriuretic peptide and patent ductus arteriosus in preterm infants. Archs Dis. Childh. *62:* 1165–1167 (1987).
14 Tulassay, T.; Rascher, W.; Hajdu, J.; Lang, R.E.; Tóth, M.; Seri, I.: Influence of dopamine on atrial natriuretic peptide level in premature infants. Acta paediat. scand. *76:* 42–46 (1987).
15 Sen, I.: Identification and solubilization of atrial natriuretic factor receptors in human placenta. Biochem. biophys. Res. Commun. *135:* 480–486 (1986).
16 Yamaji, T.; Hirai, N.; Ishibashi, M.; Takaku, F.; Yanaihara, T.; Nakayama, T.: Atrial natriuretic peptide in umbilical cord blood. Evidence for a circulating hormone in human fetus. J. clin. Endocri. Metab. *63:* 1414–1417 (1986).
17 Wei, Y.; Rodi, C.P.; Wiegand, R.C.; Needleman, L.D.; Cole, B.R.; Needleman, P.: Developmental changes in the rat atriopeptin hormonal system. J. clin. Invest. *79:* 1325–1329 (1987).
18 Aperia, A.; Broberger, O.; Elinder, G.; Herin, P.; Zetterström, R.: Postnatal development of renal function in preterm and full-term infants. Acta paediat. scand. *70:* 183–187 (1981).
19 Sulyok, E.; Németh, M.; Tényi, I.; Csaba, I.; Varga, L.; Varga, F.: Relationship between the postnatal development of the renin-angiotensin-aldosterone system and the electrolyte and acid-base status in sodium-chloride supplemented premature infants. Acta paediat. hung. *22:* 109–121 (1981).
20 Sagnella, G.A.; Markandu, N.D.; Shore, A.; MacGregor, G.A.: Effects of changes in dietary sodium intake and saline infusion on immunoreactive atrial natriuretic peptide in human plasma. Lancet *ii:* 1208–1210 (1985).

T. Tulassay, MD, First Department of Pediatrics,
Semmelweis University Medical School, Bókay J.u.53, H–1083 Budapest (Hungary)

Contr. Nephrol., vol. 67, pp. 145–148 (Karger, Basel 1988)

Endogenous Digoxin-Like Immunoreactivity in Urine of Preterm Infants with Late Hyponatremia

L. Kovács[a], *B. Lichardus*[c], *J. Birčák*[a], *M. Ponecová*[b], *A. Mašurová*[b], *J. Štefaniková*[b], *D. Bruchač*[b], *E. Sulyok*[d]

[a]First Department of Pediatrics and [b]Second Department of Obstetrics and Gynecology, Comenius University; [c]Institute of Experimental Endocrinology, CPhSci of the Slovak Academy of Sciences, Bratislava, Czechoslovakia; [d]County Children's Hospital, Pécs, Hungary

After volume expansion plasma and urine acquire enhanced natriuretic properties. There is increasing evidence that this natriuretic quality is, at least partly, mediated by an endogenous compound that interacts with the digitalis receptor Na^+,K^+-ATPase and with antidigitalis antibodies [3, 5]. Therefore, the antibodies raised against digoxin may serve as a means of detecting this proposed endogenous ligand for the Na^+,K^+-ATPase receptor [3]. For several years the presence of endogenous digoxin-like immunoreactive substance (DLIS) has been reported in human plasma and urine from individuals known never to have received the drug. The highest values of DLIS have been found in preterm infants [1, 9]. This fact is particularly intriguing because several groups of authors have shown that preterm infants have high renal losses of sodium during the first 2 weeks of life, leading to negative sodium balance and 'late' hyponatremia in many cases [6–8].

Materials and Methods

This background of experimental and clinical observations prompted us to characterize the longitudinal changes of urinary excretion of endogenous digoxin-like immunoreactive substance (UDLIS) and aldosterone (UAldo) along with sodium balance, plasma sodium concentration P_{Na}, fractional sodium excretion (FE_{Na}) and creatinine clearance measurements during the course of 'late' hyponatremia. Ten healthy preterm infant boys were selected for the study (mean gestational age 31 weeks; mean birthweight: 1,320 g).

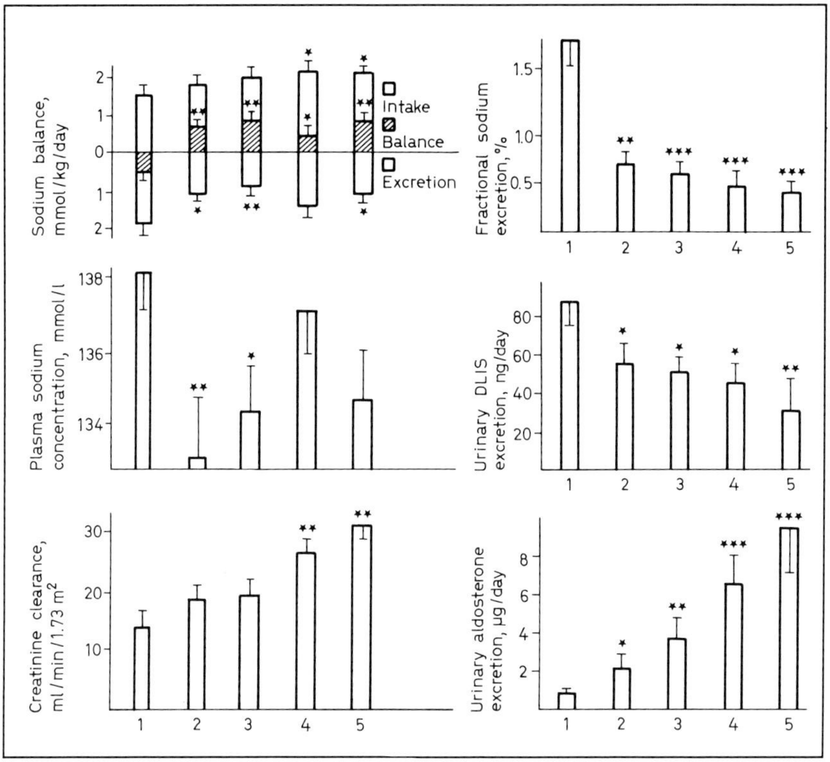

Fig. 1. Postnatal changes in sodium balance, plasma sodium concentration, creatinine clearance, fractional sodium excretion as well as urinary excretion of DLIS and aldosterone in preterm infants during the first 5 weeks of life. Asterisks indicate significant differences from data obtained in the first week: *$p < 0.05$; **$p < 0.01$; ***$p < 0.001$.

All infants were delivered vaginally and fed on pooled human milk. The parameters under investigation were determined on the seventh day and at weekly intervals thereafter up to the fifth postnatal week. UDLIS and UAldo were determined in daily urine samples by RIA (UVVVR, Prague). The study was approved by a local ethics committee and the parents gave their informed consent.

Results

The results obtained are summarized in figure 1. The most important fact is that the decreasing rate of FE_{Na} was associated with a significant decrease in UDLIS and also with an opposite change in UAldo. Moreover,

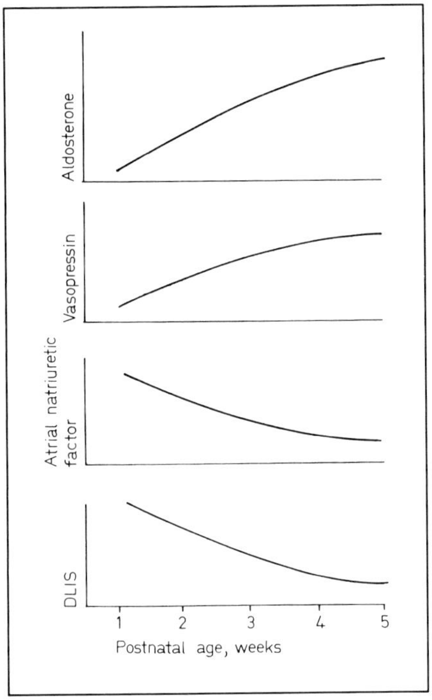

Fig. 2. Schematic summary of the changes in the activity of the major volume-regulatory systems in preterm infants during the first weeks of life [data from the present study and from references 4 and 8].

a significant positive correlation was found between UDLIS and FE_{Na} ($r = 0.53$, $p < 0.01$) as well as a negative exponential correlation between UDLIS and UAldo ($r = -0.61$, $p < 0.001$).

Discussion

It has been estimated that the postnatally occurring rapid extracellular fluid (ECF) volume contraction is higher in preterm infants in comparison to full-term babies on the basis of a larger fluid compartment in the former [2]. It is therefore expected that the sodium excretion in preterm infants would also be higher. Taking this into account, the present data (fig. 1) are consistent with the possible role of DLIS as a natriuretic hormone stimulated by neonatal hypervolemia. We suggest that the DLIS, along with the other major volume-regulatory systems, might contribute to the complex situation in which postnatal adaptation occurs [4, 6, 8]. Figure 2 shows an attractive possibility how these volume-regulatory systems may interact during the first weeks of life. It appears that postnatal natriuresis in preterm infants is promoted by both the atrial natriuretic factor and the natriuretic hormone (expressed as DLIS). With advancing ECF volume contraction the activity of both natriuretic systems decreases. Simultaneously, the activity of the two major volume-conserving systems (the renin-angiotensin-aldosterone and the vasopressin) begins to rise and after a few weeks it even predominates in the attempt to prevent excessive volume losses. The proposed natriuretic role of DLIS in preterm infants seems to be even more substantiated on the basis of our very recent finding that no decrease, but rather a slight increase in urinary DLIS excretion occurs during the first weeks of life if preterm infants are being sodium supplemented.

References

1 Ebara, H.; Suzuki, S.; Nagashima, K.; Shimano, S.; Kuroume, T.: Digoxin-like immunoreactive substances in urine and serum from preterm and term infants: relationship to renal excretion of sodium. J. Pediat. *108:* 760–763 (1986).

2 Friis-Hansen, B.: Water distribution in fetus and newborn infants. Acta paediat. scand., suppl. 305, pp. 7–11 (1983).

3 Graves, S.W.; Williams, G.H.: Endogenous digitalis-like natriuretic factors. A. Rev. Med. *38:* 433–444 (1987).

4 Kovács, L.; Sulyok, E.: Hyponatremia in preterm infants: arginine-vasopressin secretion. Archs Dis. Childh. *59:* 1107 (1984).

5 Lichardus, B.: Natriuretic hormones; in Lichardus, Schrier, Ponec, Hormonal regulation of sodium excretion, pp. 1–7 (Elsevier, Amsterdam 1980).

6 Sulyok, E.; Németh, M.; Tényi, I.; Csaba, I.F.; Győry, E.; Ertl, T.; Varga, F.: Postnatal development of renin-angiotensin-aldosterone system (RAAS) in relation to electrolyte balance in premature infants. Pediat. Res. *17:* 1013–1017 (1983).

7 Sulyok, E.; Kovács, L.; Lichardus, B.; Michajlovskij, N.; Lehotská, V.; Némethová, V.; Varga, L.; Ertl, T.: Late hyponatremia in premature infants: role of aldosterone and arginine-vasopressin. J. Pediat. *106:* 990–994 (1985).

8 Tulassay, T.; Rascher, W.; Seyberth, H.W.; Lang, R.E.; Toth, M.; Sulyok, E.: Role of atrial natriuretic peptide in sodium homeostasis in premature infants. J. Pediat. *109:* 1023–1027 (1986).

9 Valdes, E.: Endogenous digoxin-immunoreactive factor in human subjects. Fed. Proc. *44:* 2800–2805 (1985).

L. Kovács, MD, Ist Department of Pediatrics, Comenius University, Uprková 3, 811 04 Bratislava (Czechoslovakia)

Contr. Nephrol., vol. 67, pp. 149–151 (Karger, Basel 1988)

Importance of Myoglobinaemia for the Outcome of Hypoxia in Neonates

D. Boda, Z. Havass, P. Hencz, K. Bartyik, P. Temesvári, E. Eck

University Paediatric Department, Szeged, Hungary

Although a worldwide decrease of perinatal mortality was achieved by the introduction of neonatal intensive care into the clinical practice, the treatment of the adaptation disturbances of newborns to extrauterine life cannot yet be regarded as solved.

The recent publications of Kasik et al. [3] and Kojima et al. [4] reporting on myoglobinaemia in newborns with perinatal asphyxia have raised the issue that the toxic and kidney-damaging myoglobin released from the muscle into the extracellular fluid can be an additional factor responsible for the unfavourable outcome of the disease in some cases in spite of the complete intensive care.

This possibility was very important for us because elimination of the myoglobin by dialysis may explain the efficacy of peritoneal dialysis, previously introduced by us [1] in the treatment of most severe respiratory distress syndrome and other newborn cases with hypoxia.

Therefore, we found it necessary: (1) to determine the occurrence and the degree of myoglobinaemia of newborns requiring treatment in the neonatal intensive care unit (NICU), and (2) to examine experimentally the transperitoneal passage of myoglobin during peritoneal dialysis.

Material and Methods

Clinical Studies. Fifty-five newborns (mostly prematures) requiring treatment in the NICU. For laboratory tests, a RIA Myoglobin Kit was used.

Patients. Group I: 15 mild cases requiring $F_{IO_2} < 0.60$; group II: 20 severe but well-repairing cases by $F_{IO_2} < 0.60$, CPAP and/or CPEEP; group III: 20 critically severe cases needing $F_{IO_2} > 0.60$ and/or deteriorating in spite of the regular treatment in NICU.

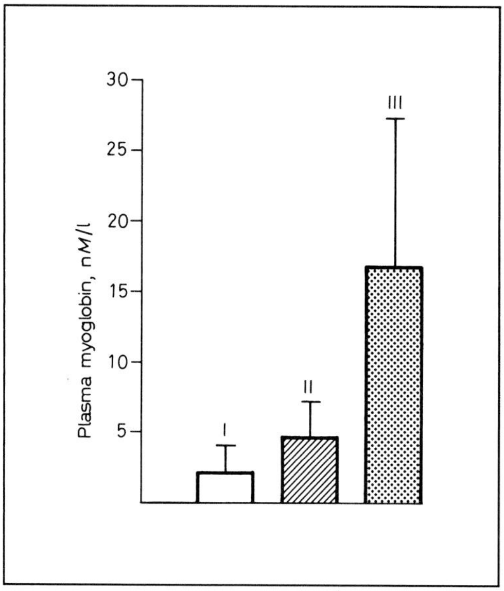

Fig. 1. Serum myoglobin values in groups I (mild cases), II (severe but rapidly recovering cases), and III (critically severe and/or rapidly deteriorating hypoxic newborns with adaptation disturbances to extrauterine life).

Experimental Studies. Four renal hilus ligated newborn piglets received i.v. myoglobin infusion, then they were dialyzed peritoneally. Myoglobin determination was performed from the sera and effluent dialyzing fluid samples.

Results

As shown in figure 1, a very low level of myoglobinaemia could also be detected in the mild newborn cases of group I. However, the most severe infants in group III were found to have highly elevated myoglobin values. These high serum myoglobin levels are comparable to those of myocardial infarction in the medicine of adults. The results of the animal experiments showed that myoglobin – being a low molecular weight protein – easily diffuses across the peritoneal membrane. The myoglobin level of the effluent dialysis fluid of the peritoneally dialyzed animals was 12.8 ± 2.6% of that of corresponding sera.

Discussion

Myoglobinaemia of newborns is one of the non-traumatic types of rhabdomyolysis. It is a consequence of the hypoxic injury of muscle cells but the myoglobin released into the extracellular fluid may act as a new damaging factor by further reducing the renal blood flow, causing acute renal failure, DIC, etc.

According to our data, severe myoglobinaemias are a relatively common finding in NICU-treated newborns. Its degree has a prognostic value. The efficacy of peritoneal dialysis has been proven previously, but it has gained a new explanation by the transperitoneal passage of the kidney-damaging myoglobin during peritoneal dialysis, so the efficacy of this treatment which was proven previously has gained a new explanation. The high level of serum myoglobin can be used as an objective finding in the indication of peritoneal dialysis. Serum myoglobin levels surpassing 10 m*M*/l seem to show the definitely high-risk cases.

According to our present results we have started a new controlled study to see the effect of peritoneal dialysis in newborn patients with most severe myoglobinaemia and hypoxia. In case of positive results the disposable dialysis sets introduced recently [2] may contribute to a broader application in clinical practice.

References

1 Boda, D.; Murányi, L.; Altorjay, I.; Veress, I.: Peritoneal dialysis in the treatment of hyaline membrane disease of newborn premature infants. Results of a controlled trial. Acta paediat. scand. *60:* 90–95 (1971).

2 Donaldson, D.C.; Spurgeon, P.; Haycock, G.B.; Chantler, C.: Peritoneal dialysis in infants. Br. med. J. *286:* 759–760 (1983).

3 Kasik, J.W.; Leuschen, M.P.; Bolam, D.M.; Nelson, R.M.: Rhabdomolysis and myoglobinaemia in neonates. Pediatrics, Springfield *76:* 255–258 (1985).

4 Kojima, T.; Kobayashi, T.; Matsuzaki, S.; Iwase, S.; Kobayashi, Y.: Effects of perinatal asphyxia and myoglobinuria on development of acute neonatal renal failure. Archs Dis. Childh. *60:* 908–912 (1985).

D. Boda, MD, University Paediatric Department, Korányi fasor 14–15,
H–6725 Szeged (Hungary)

Contr. Nephrol., vol. 67, pp. 152–154 (Karger, Basel 1988)

Indomethacin as an Inhibitor of Preterm Labor

Effect on Postnatal Renal Function[1]

A.J. van der Heijden[a], *A.P. Provoost*[b], *J. Nauta*[a], *E.D. Wolff*[a], *P.J.J. Sauer*[c]

Departments of [a]Pediatric Nephrology, [b]Pediatric Surgery, and [c]Neonatology, University Hospital Rotterdam / Sophia Children's Hospital, Rotterdam, The Netherlands

The use of indomethacin (ID) as an effective inhibitor of preterm labor was first described by Zuckerman et al. [1] in 1974. Short courses of this drug before 35 weeks of gestation are reported to be safe without any risk of side effects [2]. However, ID is known to cause a temporary decrease in renal function in neonates when used for closure of persistent ductus arteriosus. Therefore, a study on the effect of intrauterine exposure to ID was undertaken in preterm neonates after birth.

Patients and Methods

Two fatal cases were observed after long-lasting intrauterine ID exposure:

Case 1: A baby boy was born after 31 weeks of gestation. Because of preterm labor his mother was treated with ID during the last 5 weeks of pregnancy at a dosage of 400 mg daily for 4 weeks followed by 200 mg daily for 1 week. Oligohydramnios was present. The child remained anuric during his 4 days of extrauterine life and died because of respiratory and renal insufficiency.

Case 2: A girl was born after 37 weeks of gestation. Her mother was treated with ID because of preterm labor for a period of 9 weeks at a dosage of 200–400 mg daily. After delivery the patient remained anuric and was treated with peritoneal dialysis for 1 month, after which she died due to peritonitis.

Autopsy in both children revealed normal-sized kidneys. Glomeruli were small and a microcystic tubular dilatation was present.

[1] Detailed data are published separately.

Table I. Renal functional impairment in ID-exposed neonates (group I) and controls (group II): combined results of day 3, 4 and 5 of extrauterine life

GFR ml/min/kg	S_{creat} µmol/l	C_{osm} ml/min/kg	C_{H_2O} ml/min/kg	FE_{Na} %
Group I				
0.60±0.12	102±25	0.280±0.013	0.0021±0.110	4.2
Group II				
0.90±0.08	50±10	0.410±0.016	0.0183±0.180	3.0

The effect of a short course of ID during the last 48 h of pregnancy was evaluated in 9 neonates (group I) (gestational age 26–31 weeks). In these neonates inulin clearances, using continuous inulin infusion were performed to estimate the glomerular filtration rate. Serum creatinine levels were monitored. Free water clearance, osmolar clearance, urine osmolality and fractional sodium excretion were used as tubular function parameters. The neonates were compared with a control group consisting of 7 neonates without ID exposure (group II). Measurements were performed during the first 5 days of extrauterine life. Statistical analysis of the data was performed using multiple regression analysis.

Results and Discussion

Urine volume in group I was impaired, especially during the first 2 days of extrauterine life, compared to group II. Glomerular filtration rate was significantly lower, as judged from a low inulin clearance and the higher serum creatinine levels in group I. Urine osmolality was significantly higher in group I. Both osmolar and free water clearances were depressed in group I compared to group II. No significant differences in fractional sodium excretion could be observed between the two groups (table I). After about 2 weeks serum creatinine levels and fluid balance were comparable between groups I and II.

Even short courses of ID result in marked alterations in renal function during the first 5 days postnatally. ID passes the placental barrier and is slowly eliminated from the fetus as well as from the neonate, which may explain the temporary renal functional impairment. The disturbance of the balance between the high neonatal levels of angiotensin II and prostaglandins is supposed to result in a further increase in the already high renal vascular resistance in the preterm neonate. Long-lasting ID therapy can cause irreversible renal failure.

References

1 Zuckerman, H.; Reiss, U.; Ribinstein, I.: Inhibition of human premature labor by indomethacin. Obstet. Gynec., N.Y. *44:* 787–792 (1974).

2 Niebyl, J.R.; Witter, F.R.: Neonatal outcome after indomethacin treatment for preterm labor. Am. J. Obstet. Gynec. *155:* 747–749 (1986).

A.J. van der Heijden, MD, Department of Pediatric Nephrology,
University Hospital Rotterdam, Sophia Children's Hospital, Gordelweg 160,
NL–3038 GE Rotterdam (The Netherlands)

Contr. Nephrol., vol. 67, pp. 155–159 (Karger, Basel 1988)

Renal Effects of Theophylline in Very Low Birthweight Neonates

B.H. Wilkins

Department of Child Health, Southmead Hospital, Bristol, UK

Xanthines such as theophylline are widely used by neonatologists, not only for the control of apnoea of immaturity, but also increasingly to assist weaning infants from ventilator dependence [1], in which case they are often prescribed in the first week when there may be a considerable degree of renal tubular immaturity.

In adults and experimental animals xanthines act as diuretic and natriuretic agents [2] but there are few and conflicting results in premature infants [3–6]. Therefore, the acute renal effects of an acute loading intravenous dose of aminophylline have been studied in a group of preterm infants in the first week after birth.

Patients and Methods

Studies were performed in 10 newborn infants aged 1–7 days. Gestation at birth was 27.4–32.7 weeks and birthweight 1,060–1,550 g. All were receiving assisted ventilation because of severe idiopathic respiratory distress syndrome. Aminophylline (ethylene-diamine-theophylline, 81 % theophylline, in pure aqueous solution) was prescribed, while the infant was still ventilated, at a dose of 6 mg/kg birthweight given intravenously over 20 min, followed by 1 mg/kg 12-hourly. Plasma theophylline levels at the end of the loading infusion, and, subsequently, were within the therapeutic range of 28–55 μmol/l (5–10 mg/l).

Sequential randomly voided urine samples were collected for 2 h before and 4–6 h after the loading infusion and also the next day. Each sample was analysed for sodium, potassium, glucose and creatinine. A plasma sample was also assayed in each infant. No attempt was made to measure timed urine volume as this is considered impossible in small infants because of irregular and incomplete bladder emptying, and bladder catheterisation is not permitted on ethical grounds.

Sodium and potassium were measured by semiautomated flame photometry, glucose by a glucose oxidase method (Beckman glucose analyser) and creatinine by a manual end-point Jaffe method after resin adsorption of creatinine from the sample. In each baby glomerular filtration rate (GFR) was measured during the previous 24 h by inulin clearance methods [7].

For each urine sample fractional excretion of sodium (FE_{Na}) was calculated according to the formula $Na_U \times Cr_P/Na_P \times Cr_U$. On the assumption that GFR is not changing and that creatinine clearance is the same as inulin clearance, absolute sodium excretion was estimated according to the formula GFR $\times$ $Na_P \times FE_{Na}$. Fractional excretion of potassium (FE_K) and glucose ($FE_{Glucose}$) were similarly calculated. Fractional excretion of water (FE_{Water}) was calculated by Cr_P/Cr_U and urine flow rate (V) by GFR $\times$ FE_{Water}.

In order to investigate whether the loading dose of aminophylline alters GFR, in 2 cases it was given during a single injection inulin clearance experiment. Normally, plasma disappearance of inulin follows a smooth curve. Any acute perturbation in GFR will cause an acute deviation of the curve.

The study had the approval of the local medical ethics committee and informed parental consent was obtained.

Results

GFR (inulin clearance) was 0.6–1.3 ml/kg birthweight·min. Plasma levels of sodium, potassium, glucose and creatinine were normal in all cases throughout the experiment. Before aminophylline FE_{Na} was 0.9–7.7%, FE_K 4.5–52% and FE_{Water} 7.6–13%. These wide ranges presumably reflect the different ages at which the infants were studied and their different degrees of immaturity and sickness. In all cases all three quantities increased after aminophylline to peaks which occurred 0.5–4 h after the dose, and fell again to predose values by the next day (fig. 1a–c). Peak FE_{Na} was 3.2–12.7%, FE_K 12–66%, and FE_{Water} 11–32%.

Three babies had a normal $FE_{Glucose}$ of $<1\%$ and it did not rise measurably. The other 7 all had high $FE_{Glucose}$ before aminophylline

---

Fig. 1. Fractional excretion of sodium (Na), potassium (K), glucose and water (H_2O) before aminophylline, at the time of peak increase (0.5–4 h), and the next day in the study group of 10 infants. The absolute excretion values shown are calculated assuming that the GFR (inulin clearance) has not changed.

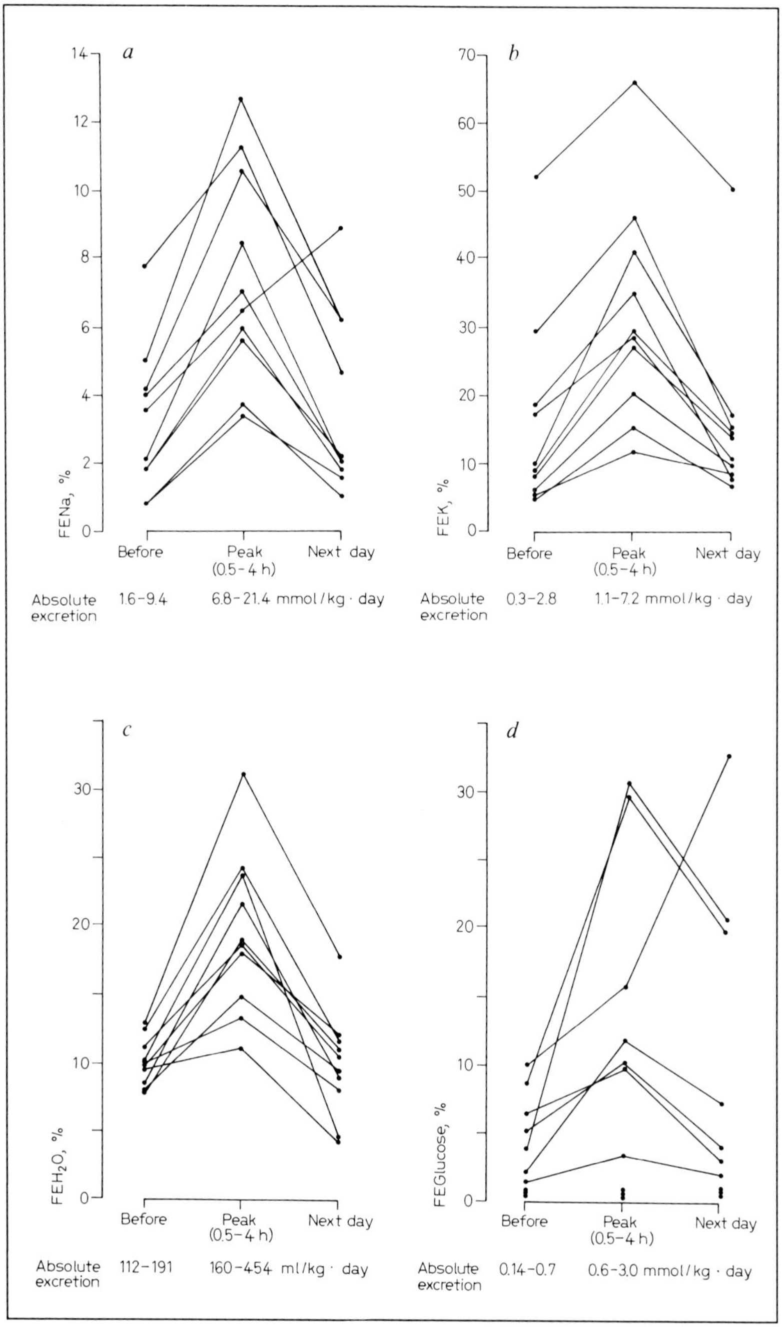
a
FENa, %
0
2
4
6
8
10
12
14
Before
Peak
(0.5-4 h)
Next day
Absolute excretion
1.6-9.4
6.8-21.4 mmol/kg · day
b
FEK, %
0
10
20
30
40
50
60
70
Before
Peak
(0.5-4 h)
Next day
Absolute excretion
0.3-2.8
1.1-7.2 mmol/kg · day
c
FEH2O, %
0
10
20
30
Before
Peak
(0.5-4 h)
Next day
Absolute excretion
112-191
160-454 ml/kg · day
d
FEGlucose, %
0
10
20
30
Before
Peak
(0.5-4 h)
Next day
Absolute excretion
0.14-0.7
0.6-3.0 mmol/kg · day

(1.5–10%) and this rose further to 6.5–31% (fig. 1d). In 3 cases this was prolonged, possibly reflecting another known side effect of aminophylline, the disturbance of cellular glucose metabolism [8].

In the 2 babies given aminophylline after inulin, there was a concomitant acute increase in the plasma disappearance of inulin lasting approximately 1 h. This must reflect an acute increase in GFR, because it is unlikely that there is a sudden change in inulin distribution space. The complexity of inulin disappearance curves, however, prevents quantification of this increase, either in magnitude or duration.

Discussion

In this study systemic administration of theophylline causes a rapid increase in the renal excretion of sodium, potassium, water and glucose in a group of babies who already, because of sickness and immaturity, have a high renal sodium loss. This change, however, only lasts a few hours despite continuing therapeutic plasma levels of theophylline.

Xanthines are inhibitors of intracellular phosphodiesterase and might be expected therefore to decrease tubular reabsorption of sodium, potassium and water although the increased glucose excretion is more difficult to explain. Xanthines, especially theophylline, are also inhibitors of adenosine [9]. This has been postulated to be the cause of the xanthine-associated increase in GFR in newborn rabbits given similar doses of aminophylline as in this study [10].

The apparent acute increase in GFR detected in 2 cases in this study suggests that the observed tubular effects may in part be due to the fact that the tubules are temporarily presented with an increased load of filtrate. Because the tubules are already compromised in their ability to reabsorb sodium (and maybe potassium and glucose) then it is not surprising that the fractional reabsorption falls. The absolute magnitude of the extra sodium and potassium excreted cannot be estimated using the methods of this study.

It is likely that theophylline is exerting a vasoactive effect on the small blood vessels of the kidney. If this extends beyond the glomerulus to other systems, e.g. the cerebral circulation, where small acute changes in perfusion may be harmful, then rapid loading doses of aminophylline may not be appropriate.

References

1 Elias-Jones, A.; Dhillon, S.; Greenough, A.: The efficacy of oral theophylline in ventilated premature infants. Early hum. Dev. *12:* 9–14 (1985).

2 Rall, T.: Central nervous system stimulants: the methylxanthines; in Goodman, Gilman, The pharmacologic basis of therapeutics; 7th ed., pp. 589–603 (Macmillan, New York 1985).

3 Harkavy, K.; Scanlon, J.; Jose, P.: The effects of theophylline on renal function in the premature newborn. Biol. Neonate *35:* 126–130 (1979).

4 Howell, J.; Clozel, M.; Aranda, J.: Adverse effects of caffeine and theophylline in the newborn Infant. Semin. Perinatol. *5:* 359–369 (1981).

5 Rothberg, A.; Marks, K.; Ward, R.: The metabolic effects of caffeine in the newborn infant. Pediat. Pharmacol. *1:* 181–186 (1981).

6 Zakauddin, S.; Leake, R.; Trygstad, C.: Renal effects of theophylline in the premature infant. Clin. Res. *24:* 196 (1976).

7 Coulthard, M.: Comparison of methods of measuring renal function in preterm babies using inulin. J. Pediat. *102:* 923–930 (1983).

8 Srinavasan, G.; Singh, J.; Cattamanchi, G.; Yeh, T.; Pildes, R.: Plasma glucose changes in preterm infants during oral theophylline therapy. J. Pediat. *103:* 473–476 (1983).

9 Persson, C.; Andersson, K.; Kjellin, G.: Effects of enprofylline and theophylline may show the role of adenosine. Life Sci. *38:* 1057–1072 (1986).

10 Gouyon, J.-B.; Guignard, J.-P.: Renal effects of theophylline and caffeine in newborn rabbits. Pediat. Res. *21:* 615–618 (1987).

Barry Wilkins, MD, Department of Child Health, Southmead Hospital, Southmead Road, Bristol BS10 5 NB (UK)

Contr. Nephrol., vol. 67, pp. 160–165 (Karger, Basel 1988)

Renal Handling of Phosphate in Term and Preterm Infants

K. Scholz, R. Burghard, J.U. Leititis, M. Brandis

Department of Pediatrics, Philipps-University, Marburg/Lahn, FRG

Preterm infants are at greater risk for developing rickets during the first months of life. Vitamin D and calcium deficiency as well as inadequate intake of phosphate have been implicated as causative factors for the bone demineralization. Immature kidney function may also partly be responsible for the less-efficient control of calcium and phosphate homeostasis in preterm as compared to term infants.

The investigations of other authors suggested that tubular function in the newborn is less mature than glomerular function, which would result in potential glomerulotubular imbalance and insufficient reabsorption of filtered substances like glucose, bicarbonate, and *p*-aminohippurate [1–3]. Earlier studies of tubular reabsorption of phosphate gave conflicting results [2, 4–6].

The aim of our investigation was to study renal handling of inorganic phosphate and the regulative mechanisms responsible for the maintenance of phosphate homeostasis during the perinatal period.

Materials and Methods

Physiologically stable newborn infants (n = 37) with gestational ages (GA) of 30–42 weeks were studied during the first week of life (2–6 days). The estimate of GA was made by clinical data and the Dubowitz score. Data were grouped for presentation according to GA in the following manner: 30–34 weeks, 35–37 weeks and 38–42 weeks.

Infants with a GA of 30–34 weeks were studied during the second week after birth (n = 6) and during the 3rd to 6th weeks of life (n = 10). All newborns were clinically stable, none of them had a respiratory distress syndrome, neonatal asphyxia or septicemia. None required artificial respiration. Drugs with suspected effects on renal function were not given prior to or during the investigation.

Short-term clearances as described by Arant [4] were performed in the morning. Blood samples were taken 30 min after the collection period had started, the plasma was separated and stored at −28 °C. Inorganic phosphate was assayed using a modification of the phosphomolybdate method. Creatinine was analyzed by kinetic approach to the Jaffe reaction. PTH and cAMP were determined by radioimmunoassay (Ria-Mat, PTH-MM from Mallinckrodt and RIA Kit by Amersham).

Medians and the 95% confidence limits were calculated and nonparametric tests were used to calculate the significance of differences. The endogenous creatinine clearance was taken as the estimate of GFR. Using standard clearance techniques the following parameters of renal phosphate handling were calculated: Cp = phosphate clearance; Tp = net phosphate reabsorption; Tp/CCrea = fractional phosphate reabsorption; GFR × P = glomerular-filtered P; FEp = fractional p excretion; UpV = absolute p excretion.

All values were corrected to body surface area.

Results

Tubular Reabsorption of Phosphate

Results of renal phosphate handling in the first postnatal week are demonstrated in table I. No differences between preterm and term infants could be confirmed concerning the UpV, the FEp, Cp, Tp and Tp/CCrea.

Table I. Parameters of glomerulotubular phosphate transport in term and preterm infants in relation to gestational age

	Group A 30–34 weeks	Group B 35–37 weeks	Group C 38–42 weeks	p
UpV, mmol/BW/24 h	0.23 (0.02–0.49)	0.2 (0.02–0.48)	0.14 (0.04–0.51)	0.76
Cp, ml/min × 1.73 m^2	1.5 (0.1–3.6)	0.9 (0.2–2.9)	1.1 (0.3–4.4)	0.91
Tp, μmol/min × 1.73 m^2	23.3 (10.3–40.5)	47.8 (19.6–56.5)	53.3 (41.9–66.3)	0.08
Tp/CCrea, μmol/ml	1.75 (1.6–2.0)	2.0 (1.5–2.4)	2.0 (1.8–2.2)	0.73
FEp, %	6.2 (1.1–20.5)	6.6 (0.8–12.5)	3.0 (0.9–18.5)	0.49
n	10	14	12	

p = Kruskal-Wallis test.

Table II. Parameters of glomerulotubular phosphate transport in preterm infants in relation to postnatal age

	Weeks of life			p
	1 week	2 weeks	3–6 weeks	
UpV, mmol/BW/24 h	0.23 (0.02–0.49)	0.18 (0.09–0.57)	0.46 (0.31–0.92)	0.26
Cp, ml/min × 1.73 m^2	1.5 (0.1–3.6)	1.7 (0.6–2.7)	3.1 (2.2–5.4)	0.18
Tp, μmol/min × 1.73 m^2	23.3 (10.3–40.5)	47.3 (30.5–55.4)	51.8 (41.4–70.0)	0.13
Tp/GFR, μmol/ml	1.75 (1.6–2.0)	1.95 (1.6–2.1)	2.1 (1.9–2.2)	0.26
FEp, %	6.3 (1.1–20.5)	7.4 (3.5–12.4)	9.7 (7.7–18.2)	0.51
n	10	6	9	

p = Kruskal-Wallis test.

During the first 6 weeks of life no significant changes of phosphate reabsorption could be demonstrated (table II). All study groups exhibited a positive phosphate balance even during the first week of life. Under these conditions the plasma phosphate (Pp) levels were not different irrespective of gestational or conceptional age.

In order to evaluate the renal mechanisms which could lead to a constant level of Pp in newborn infants, Pp was correlated to the various renal parameters investigated. A significant correlation was found only between Pp and Tp/GFR, as shown in figure 1. The correlation is in the same order of magnitude compared with those for infants aged from 0.5 to 12 months obtained by other investigators [5].

Increased filtered load (GFR × P) was accompanied by a simultaneous increase of tubular net reabsorption (Tp) regardless of gestational or conceptional age (fig. 2). Consequently, a strong correlation exists between GFR × P and Tp in all groups with identical regression lines.

PTH serum levels and urinary cAMP excretion in preterm infants exhibited significantly higher PTH values than in the term newborns. PTH was inversely correlated to GA (fig. 3). Excretion of cAMP is depicted in figure 4. PTH serum levels were neither correlated with tubular P reabsorption nor with cAMP excretion.

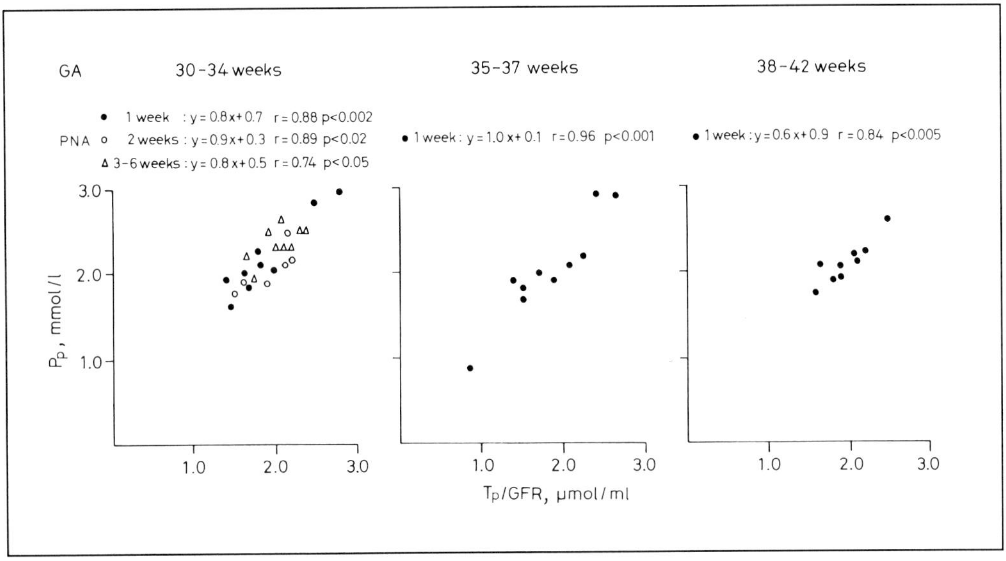

Fig. 1. Correlation between Pp and Tp/GFR.

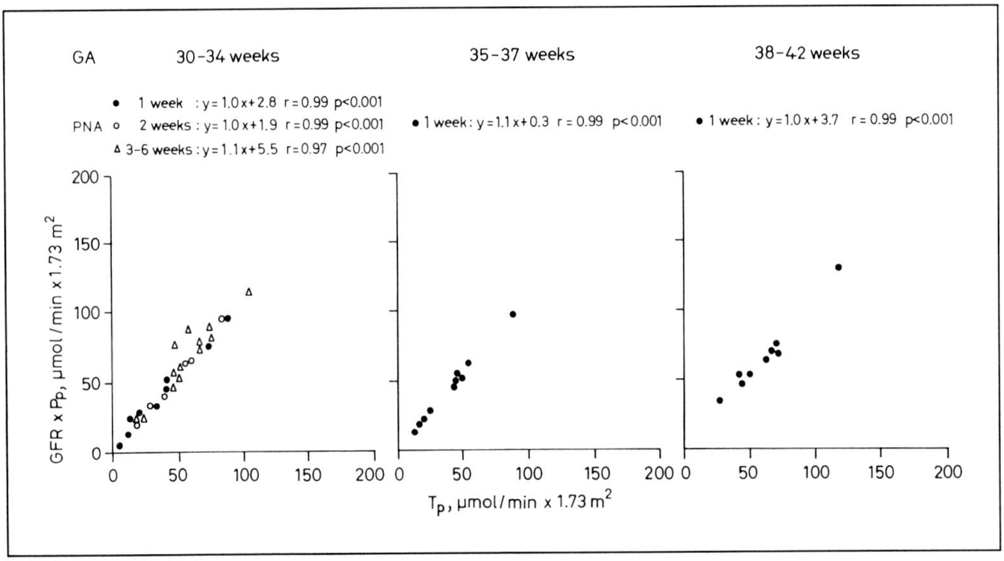

Fig. 2. Correlation between GFR × Pp and Tp.

3 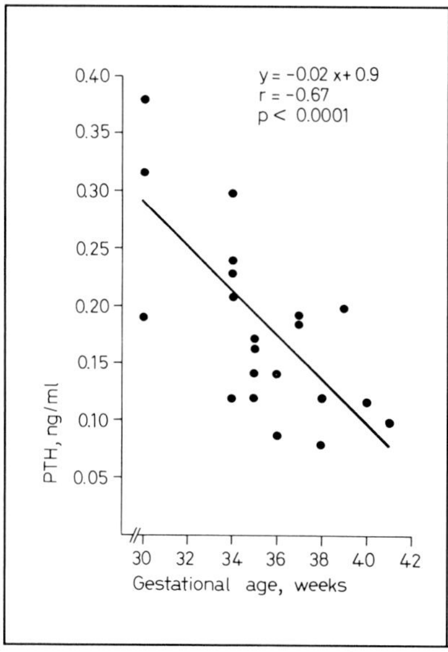

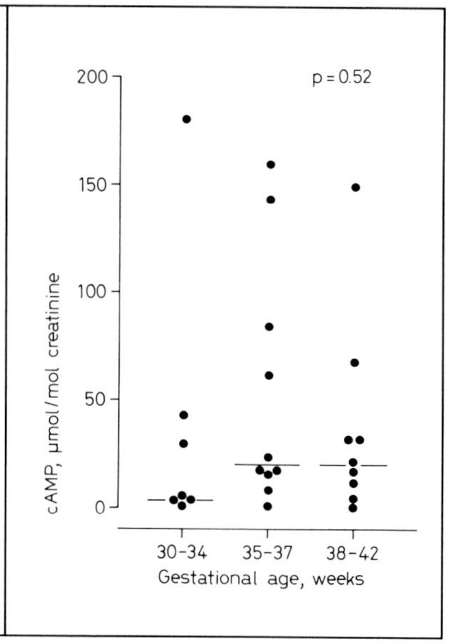 4

Fig. 3. Correlation between PTH and gestational age.
Fig. 4. Excretion of cAMP in preterm and term newborns.

Discussion

Tubular reabsorption has been reported to be decreased in preterm infants during the first postnatal week [6]. The results of our investigation, however, show that there is no difference between preterm and term infants concerning the tubular handling of phosphate during the first 6 weeks of life. Glomerular-tubular balance for phosphate seems to be established even in preterm infants immediately after birth under physiological conditions: increased filtered load is accompanied by a simultaneous increase of tubular net reabsorption. Therefore, immature kidney function is not responsible for rickets in prematurity.

In accordance with other studies [5], we found a close correlation between plasma phosphate and fractional phosphate reabsorption in preterm as well as in term newborn infants. Our data therefore support the hypothesis of Brodehl et al. [5] that the fractional phosphate reabsorption is the major determinant for the maintenance of high plasma phosphate level in newborns.

Another factor to be discussed is the decreased responsiveness of the tubules to PTH. The results of our investigation show that tubular phosphate reabsorption in newborns is not dependent on PTH. Higher PTH levels in preterm infants do not result in stimulation of cAMP.

On the basis of our studies, it cannot be discussed whether another homeostatic factor might be effective in regulation of the tubular phosphate reabsorption as speculated by Brodehl et al. [5].

References

1 Tudvad, F.; Vesterdal, J.: The maximal tubular transfer of glucose and para-aminohippurate in premature infants. Acta paediat. *42:* 337 (1953).

2 Brodehl, J.; Gellissen, K.: Endogenous renal transport of free aminoacid in infancy and childhood. Pediatrics, Springfield *42:* 395 (1968).

3 Barnett, H.L.: Kidney function in young infants. Pediatrics, Springfield *5:* 171 (1950).

4 Arant, B.S.: Developmental patterns of renal functional maturation compared in the human neonate. Pediatrics, Springfield *92:* 705 (1978).

5 Brodehl, J.; Gellissen, K.; Weber, H.P.: Postnatal development of tubular phosphate reabsorption. Clin. Nephrol. *17:* 163 (1982).

6 Karlen, J.; Aperia, A.; Zetterström, R.: Renal excretion of calcium and phosphate in preterm and term infants. Pediatrics, Springfield *106:* 814 (1985).

Kathrin Scholz, MD, Department of Pediatrics, Phillips-University,
Deutschhausstrasse 12, D-3550 Marburg/Lahn (FRG)

Contr. Nephrol., vol. 67, pp. 166–170 (Karger, Basel 1988)

Renal Tubular Handling of Lactate in Neonates

Index of Tubulo-Vascular Adaptation[1]

L.B. Zimmerhackl, S. Blum, M. Kramer, P. Küster, P. Rausch, K. Steidel, R. Burghard, J.U. Leititis, M. Brandis

Department of Pediatrics, Philipps-University, Marburg/Lahn, FRG

The kidney develops throughout the fetal life. The development of the metanephric nephron is initiated in the mid to late period of pregnancy. *Structurally*, the development of the nephrons are completed at about 34 weeks of gestation [1]. After birth, however, the tubules continue to grow. *Morphologically*, the renal cortex increases in volume and the medullary thick ascending loop of Henle elongates and increases in relative size [8, 9]. *Functionally*, renal plasma perfusion and glomerular filtration rate are extremely low during fetal life as a consequence of high intrarenal resistance [3, 8]. Postnatally, the renal plasma flow and glomerular filtration rate augment dramatically while renal resistance diminishes. Between tubular transport activity and renal perfusion a vascular-tubular balance has to develop [1, 8]. In states of imbalance, that is, disparity between tubular transport and vascular energy supply, the kidney uses for compensation anaerobic glycolysis with intrarenal lactate production [2, 7, 10]. Figure 1 summarizes the key metabolic pathways of lactate in the kidney. Lactate is freely filtered in the glomerulus and almost completely reabsorbed in the proximal tubule via an active sodium-coupled transport system [5]. Under the conditions of intrarenal energy deficiency the thick ascending loop of Henle and distal tubular structures can produce ATP with anaerobic glycolysis [6]. Increased glycolysis leads to increased intracellular lactate concentration and passive diffusion into the tubular lumen

[1] Supported by Kempkes Stiftung, Marburg.

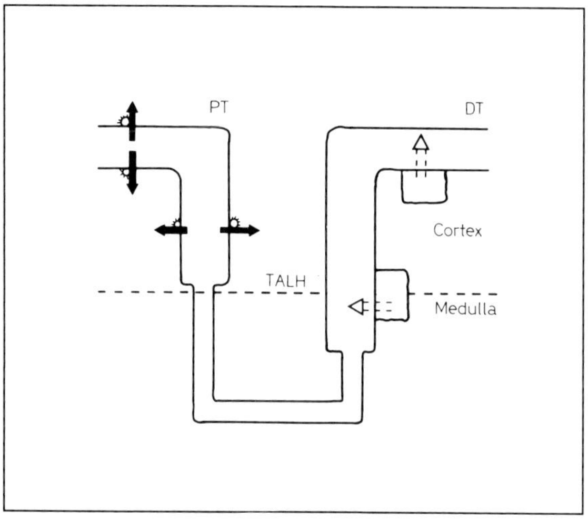

Fig. 1. Schematic presentation of intrarenal lactate metabolism. Proximal tubular active transport of lactate is indicated by the solid arrows; passive cellular to luminal diffusion of lactate is indicated by the broken arrows. PT = Proximal tubule; TALH = thick ascending loop of Henle; DT = distal tubule.

[5]. Since no reabsorptive transport systems for weak organic acids are described further downstream, lactate is excreted quantitatively. Elevated excretion of lactate can be caused by decreased proximal tubular reabsorption (maturational cause), or as a consequence of increased intrarenal production at more distal sites (adaptational cause). In the present study we determined renal lactate handling in neonates in order to evaluate maturational or adaptational factors. Excretion of N-acetyl-glucosaminidase activity, a lysosomal enzyme of the proximal tubule, was determined to assess proximal tubular maturation.

Patients and Methods

Patients

Eighty-six neonates with gestational ages between 33 and 42 weeks were investigated for renal tubular transport of lactate at days 1, 2–4 and 5–10. Neonates with septicemia and renal insufficiency were excluded. Influence of gestational age was investigated comparing preterm infants with gestational age from 33 to 37 weeks with neonates with gestational age ranging from 38 to 42 weeks.

Table I. Summary of the data

	Day 1	Days 2–4	Days 5–10	p
Serum lactate, mmol/l	6.6	5.8	5.1	n.s.
Filtered load of lactic acid, µmol/min × 1.73 m^2	77.1	66.5	124.4	n.s.
Urine lactate, mmol/l	1.72	1.27	0.27	<0.05
Urine lactate/creatinine, mmol/mmol	0.84	0.55	0.08	<0.05
Lactate excretion per body weight, µmol/min × kg	34.6	12.5	5.8	<0.05
FE lactate, %	1.38	0.64	0.37	<0.04
N-acetyl-glucosaminidase in urine, kU/l	2.75	5.25	6.0	n.s.

Median values. n.s. = Not significant.

Methods

Blood was withdrawn from peripheral veins and immediately deproteinized. Urine was sampled by timed collection. Lactate concentration was determined in serum and urine by cation-exchange high-performance liquid chromatography (HPLC) as described elsewhere [11]. Glomerular filtration rate was assessed with creatinine clearance and corrected for body surface area.

Statistics

Nonparametric tests were used. Median and quartile were calculated and group comparison for statistical significance was made using the Kruskal-Wallis or Wilcoxon rank sum test, as appropriate.

Results

Table I summarizes the data. Serum lactate concentration and filtered lactate load did not change within the first 10 postnatal days. Urinary concentration of lactate, urinary lactate to creatinine ratio and total lactate excretion corrected for body surface area or body weight decreased significantly, respectively. This resulted in a significant fall in fractional excretion of lactic acid.

N-acetyl-glucosaminidase concentration in urine remained unchanged within the postnatal period studied. There was no significant influence of gestational age by comparison of the neonates 33–37 weeks of gestation with those of 38–42 weeks.

Discussion

Postnatal adaptation is a demanding process to the maturing kidney. Progressive amounts of solute are filtered and tubular reabsorption is increased. The successful process of tubulo-vascular balance is, however, crucial for kidney function. The kidney is an organ which can metabolize and produce lactate. Regarding glucose and lactate metabolism there is a marked heterogeneity. Proximal tubular structures have marked enzyme activities for gluconeogenesis. In contrast, the thick ascending loop of Henle and distal tubular structures are equipped with glycolytic enzymes [2]. However, direct tubular transport characteristics cannot be determined with the present methods. 'Black box' analysis is the only suitable investigative way in humans. In order to differentiate between proximal tubular lactate reabsorption and more distal lactate production, we used N-acetyl-glucosaminidase, an enzyme of lysosomes predominantly of the proximal tubule. No ideal marker for proximal tubular maturation has been found, however [4]. In comparison with experimental data, the proximal tubule does increase in size and volume [9]. Therefore, NAG excretion does reflect one proximal tubular marker. There may be, however, heterogeneity regarding different transport systems as discussed by others [4].

In summary, fractional excretion of lactic acid decreases significantly after birth, while proximal tubular parameters remain unaltered after birth. In the range of neonates studied gestational age was not of significant influence. Therefore, the postnatal decrease in fractional lactate excretion may reflect decreased intrarenal lactate production and not increased proximal tubular lactate reabsorption. Decreased intrarenal lactate secretion would then indicate progressive postnatal tubulo-vascular adaptation.

References

1 Arant, B.S.: Postnatal development of renal function during the first year of life. Pediat. Nephrol *1:* 308–313 (1987).

2 Guder, W.G.; Ross, B.D.: Enzyme distribution along the nephron. Kidney int. *26:* 101–111 (1984).

3 Jose, P.A.; Logan, A.G.; Slotkoff, L.M.; Lillienfield, L.S.; Calcagno, P.L.; Eisner, G.M.: Intrarenal blood flow distribution in canine puppies. Pediat. Res. *5:* 335 (1971).

4 Leititis, J.U.; Burghard, R.; Gordjani, N.; Kaethner, T.; Brandis, M.: Entwicklungsphysiologische Aspekte der Volumen- und Natriumregulation bei Frühgeborenen und reifen Neugeborenen. Mschr. Kinderheilk. *135:* 3–9 (1987).

5 Murer, H.; Barac-Nieto, M.; Ullrich, K.J.; Kinne, R.: Renal transport of lactate; in Greger, Oberleithner, Silbernagl, Renal tubular transport, pp. 210–223 (Springer, Heidelberg 1982).
6 Ross, B.D.; Espinal, J.; Silva, P.: Glucose metabolism in renal tubular function. Kidney int. *29:* 54–67 (1986).
7 Schaerer, K.: Die Laktatazidose; Habschr. Universität Heidelberg (1969).
8 Spitzer, A.: Renal physiology and functional development; in Edelmann, Pediatric kidney disease, vol. I, pp. 25–127 (Little, Brown, Boston 1978).
9 Spitzer, A.; Brandis, M.: Functional and morphologic maturation of the superficial nephron. Relationship to total kidney function. J. clin. Invest. *53:* 279 (1974).
10 Wirthensohn, G.; Guder, W.G.: Renal substrate metabolism. Physiol. Rev. *66:* 469–497 (1986).
11 Zimmerhackl, L.B.; Kramer, M.; Blum, S.; Küster, P.; Rausch, P.; Steidel, K.; Brandis, M.: Eine einfache HPLC-Methode zur Bestimmung von Milchsäure in Plasma und Urin. Lab. Med. *11:* 171 (1987).

Bernd Zimmerhackl, MD, Department of Pediatrics, Philipps-University, Deutschhausstrasse 12, D-3550 Marburg/Lahn (FRG)

Contr. Nephrol., vol. 67, pp. 171–174 (Karger, Basel 1988)

Prenatal Diagnosis of Urological Abnormalities

T.M. Barratt, M.J. Dillon, P.G. Duffy, J. Fay, I. Gordon, P.G. Ransley

Department of Paediatric Nephrology, Institute of Child Health,
Hospital for Sick Children, London, UK

The widespread use of routine ultrasound in pregnancy has led to the antenatal diagnosis of urological abnormality in a substantial number of babies [1]. They comprise a wide variety of anatomical lesions with renal function ranging from normality to complete anuria, and present a new range of problems in diagnosis, assessment, management and treatment to the obstetrician and paediatric radiologist, nephrologist, and urologist.

Incidence

The incidence of urological abnormality has not been defined, and will depend upon the population of pregnant women screened, the timing, frequency and assiduity of the ultrasonography, and the experience of the operator.

From 1979 to 1986 inclusive, 180 infants with an antenatal diagnosis of urological abnormality were referred to the Department of Urology and Renal Unit of the Hospital for Sick Children, London (table I). The referral rate appears to have stabilised at 50–60 per annum, and now accounts for 60–70% of infants admitted with congenital urological disorders. The outcome for these infants was good, with only 9 having died or being left with chronic renal failure (CRF: plasma creatinine concentration (PCr) > 150 μmol/l at 1 year of age). Surprisingly, most infants with CRF due to uro-

Table I. Infants referred with urological abnormality, Hospital for Sick Children, London, 1979–1986; (n) = antenatal diagnosis

	PCr at 1 year		
	<75 μmol/l	75–150 μmol/l	>150 μmol/l
1979	24 (2)	1	7
1980	35 (1)	4	6
1981	32 (3)	4	4
1982	52 (11)	2 (1)	4 (1)
1983	73 (24)	6 (1)	6 (2)
1984	64 (20)	5	5 (1)
1985	87 (59)	6	8 (2)
1986	65 (46)	5 (3)	9 (3)
Total	432 (166)	33 (5)	49 (9)

Table II. Antenatal diagnosis of urological abnormality, Hospital for Sick Children, London, 1979–1986

	PCr at 1 year		
	<75 μmol/l	75–150 μmol/l	>150 μmol/l
Pelvi-ureteric junction 'obstruction'			
U/L	65		
B/L	14	1	
Multicystic kidney, U/L	32		
Cystic/dysplastic kidney, B/L	6	1	4
Ureterovesical junction 'obstruction'			
U/L	5		
B/L	3		
Ureterocele, U/L	8		
Vesicoureteric reflux			
U/L	8		
B/L	9	1	
Posterior urethral valve	9	2	2
Prune belly syndrome	3		1
Miscellaneous	4		2
Total	166	5	9

U/L = Unilateral; B/L = bilateral.

logical abnormality still appear to be escaping the net of antenatal diagnosis, with only 5 of 17 (29%) of such cases in the last 2 years having been detected before birth.

Pathology

The spectrum of abnormality differs according to the source of cases: reports from obstetric units include a greater proportion of severely affected infants with renal agenesis, oligohydramnios, and pulmonary hypoplasia, who do not survive, whereas those emanating from paediatric urological departments are dominated by mildly affected infants, many of whom remain asymptomatic and would not otherwise have come to medical attention (table II).

The most common abnormality in infants with an antenatal diagnosis of urological abnormality referred to the Hospital for Sick Children, London, is a simple hydronephrosis with normal lower ureter, bladder and outflow tract. It may be unilateral or bilateral; the renal pelvis may be hugely dilated or only transiently distended; renal function may be normal or severely impaired. The fundamental abnormality is assumed to be an impediment to the passage of urine through the pelvi-ureteric junction, but dilatation of the renal pelvis does not of itself necessarily imply significant obstruction, and rigorous urodynamic criteria for the diagnosis of urinary obstruction in these infants have not been established, nor is it clear which patients require corrective surgery and which can be managed conservatively [2].

Discussion

Several questions remain open for discussion. Can intrauterine urinary tract obstruction resolve completely, either without residual abnormality, or with renal dysplasia together with, in the prune belly syndrome, dilatation of the ureters and bladder, absent abdominal musculature, and undescended testicles [3]? Can the severity of renal abnormality be judged from amniotic fluid volume and composition? Is there a case for intrauterine surgery in selected cases to preserve renal function [4]? What are the appropriate protocols for post-natal assessment? What are the criteria for the diagnosis of obstruction, and which cases require surgery?

References

1 Woodward, J.R.; Parrott, J.S.: Neonatal urology; in Holliday, Barratt, Vernier, Pediatric nephrology, pp. 945–962 (Williams & Wilkins, Baltimore 1987).
2 Manzoni, A.; Ransley, P.G.: Postnatal management of UPJ obstruction detected antenatally. Dialog pediat. Urol. *8:* 6–7 (1985).
3 Pagon, R.A.; Smith, D.W.; Shepherd, T.M.: Urethral obstruction malformation complex. A cause of abdominal muscle deficiency and the 'prune-belly'. J. Pediat. *94:* 900–911 (1979).
4 Harrison, M.R.; Filly, R.A.; Parer, J.T.; et al.: Fetal surgery for congenital hydronephrosis. New Engl. J. Med. *306:* 591–596 (1982).

Prof. T.M. Barratt, Department of Paediatric Nephrology,
Institute of Child Health, 30 Guilford Street, London WCIN 1EH (UK)

Contr. Nephrol., vol. 67, pp. 175–182 (Karger, Basel 1988)

Role of Imaging in the Postnatal Period following Antenatal Diagnosis

I. Gordon

Department of Nephrology, Institute of Child Health, Hospital for Sick Children, London, UK

The role of the paediatric radiologist in the clinical situation of antenatally diagnosed renal abnormalities is to confirm the suspected abnormality acting as part of a nephro-urological team. The questions posed by the clinical team are to define the kidneys and collecting systems in great detail as well as defining individual renal function. This demands the use of ultrasound (US), micturating cystourethrography (MCU), Tc99m dimercaptosuccinic acid (DMSA) or Tc99m DTPA scans and intravenous urography (i.v.u.).

Ultrasound

The first examination should always be an US. This examination must include detailed analysis of the bladder wall as well as searching for dilated ureters (fig. 1). The normal appearance of the kidneys include an echolucent medulla and a relatively echo-dense cortex. A small kidney with loss of the cortico-medullary differentiation suggests renal dysplasia which may be associated with cysts to varying degrees (fig. 2a, b). The presence of dilatation of the collecting system must raise the possibility of obstruction, if the dilatation is bilateral then bladder outlet pathology must be excluded. However, vesico-ureteric reflux (VUR) may give a very similar appearance (fig. 3a, b).

In polycystic disease it is not always possible to distinguish the recessive from the dominant form since both the fetus and infant have been shown to have the typical appearance of the recessive form early on but on follow-up in later infancy and childhood the appearance has changed to that of the dominant polycystic type.

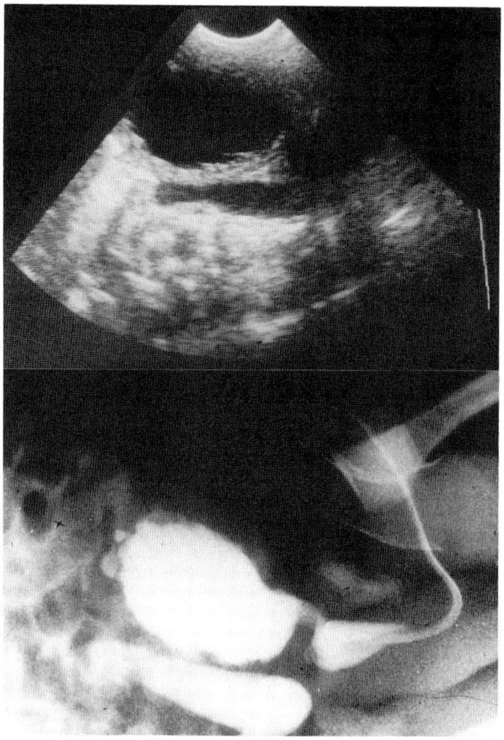

Fig. 1. The US top image demonstrates the full bladder with the dilated ureter behind the bladder. The dilated posterior urethra is clearly seen as is the bladder neck hypertrophy. The orientation of the MCU is the same as the bladder, correlation between the anatomical detail provided by these two examinations. The MCU further demonstrates the posterior urethral valve as the cause of the dilated posterior urethra.

The US can be carried out on the first day of life when the antenatal diagnosis was that of bilateral hydronephrosis; however, in the case of suspected unilateral hydronephrosis then the US should be done on the 3rd post-natal day in order not to miss significant renal pelvic dilatation.

Micturating Cystourethrogram

The MCU gives invaluable information in the presence of bilateral hydronephrosis. This includes evidence of bladder outflow obstruction

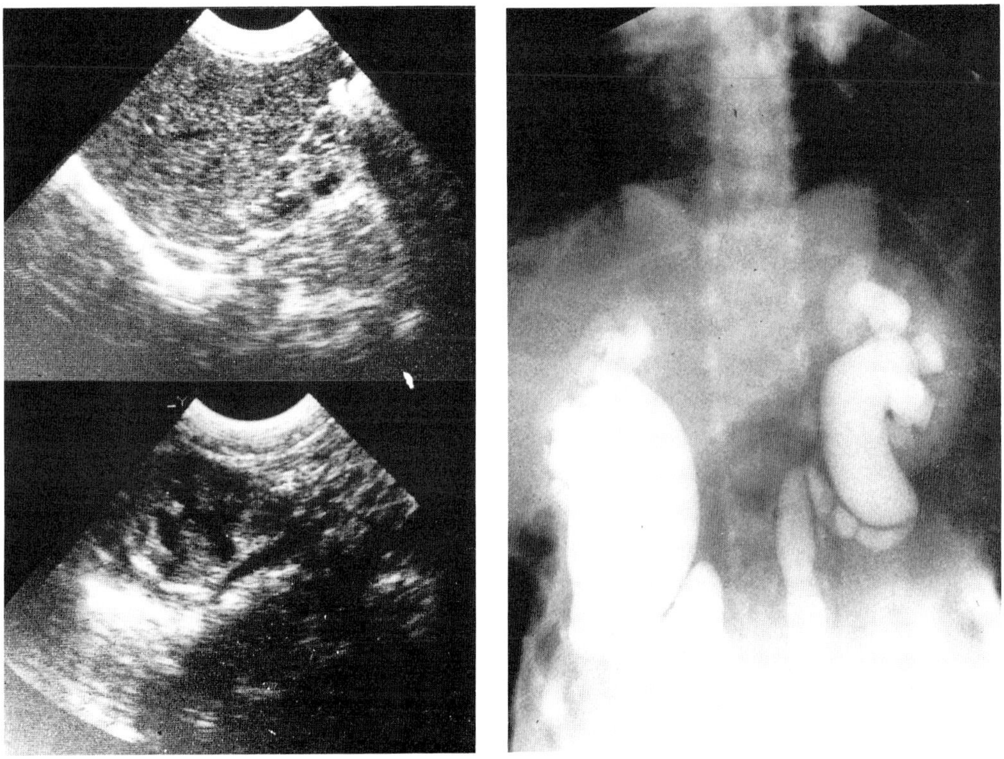

a *b*

Fig. 2. a The two US images show that the kidney has lost its cortico-medullary differentiation and that the entire kidney is bright compared to the liver. Minimal dilatation of the collecting system can be seen on the lower image. The kidney is also noted to be small. *b* The MCU shows bilateral VUR, the calyces are abnormal as is the renal pelvis, and these appearances strongly support the ultrasound suggestion that renal dysplasia is present.

and/or VUR (fig. 1, 2b). In the presence of VUR, contrast outlining the calyces may suggest the presence of renal dysplasia (fig. 2b, 3b). The combination of US and MCU allows adequate evaluation of all kidneys and collecting systems shortly after birth and permits appropriate management to be instituted immediately. All further imaging may be planned when the neonate is stable and full discussion of the clinical, laboratory and radiological findings have been evaluated by the clinical team. This is even more important if there are other congenital abnormalities unrelated to the urinary tract.

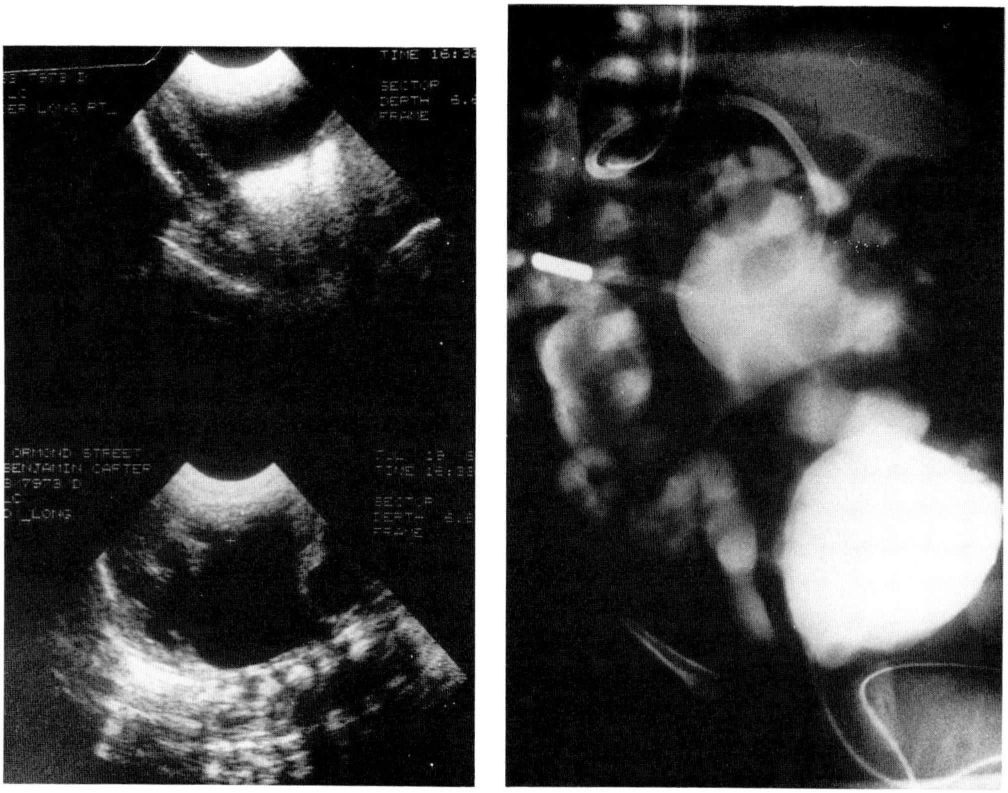

a b

Fig. 3. a The US demonstrates significant hydronephrosis involving the calyces and pelvis of the kidney (lower image). The bladder is full of urine and the dilated ureter can be seen behind the bladder (upper image). *b* The MCU clearly demonstrates bilateral VUR. No bladder neck hypertrophy could be demonstrated and the urethra appeared normal (not shown). This child had bilateral vesico-ureteric reflux with no evidence of any bladder outflow obstruction, yet the ultrasound appearances of the kidneys and ureters strongly suggested an outflow obstruction. This demonstrates the need for the MCU early on.

Further imaging is usually related to obtaining functional information about each kidney. This requires a radioisotope examination. Transitional nephrology must be borne in mind since this governs the way both Tc99m DTPA and contrast for an i.v.u. as well as Tc99m DMSA are handled by the kidney.

Dynamic Renal Isotope Scan

Tc99m DTPA as well as the contrast used for i.v.u. are pure glomerular filtrates being neither reabsorbed nor excreted by the tubules. Since the neonate has a relatively large extracellular fluid space and a low glomerular filtration rate, the normal Tc99m DTPA is characterised by a high background count rate, a rapid renal transit time with poor delineation of the kidney. The half-life in a normal full-term neonate of the DTPA or contrast for i.v.u. is approximatey 3 times longer than that of adolescence. These physiological reasons suggest that one should wait as long as possible before carrying out a Tc99m DTPA scan (fig. 4). In the presence of one normal kidney then unilateral hydronephrosis may be further investigated at the end of the neonatal period. When posterior urethral valves are present, then the first Tc99m DTPA scan is carried out 2 weeks after valve ablation, and follow-up scans are done regularly at 3, 12 and 24 months.

Intravenous Urogram

The i.v.u. is reserved to answer specific problems. The most common indication is to provide certain anatomical details essential for further management. This occurs most frequently when a ureterocoele has been identified but the US had failed to diagnose from which kidney this arises. In the presence of gross bilateral hydronephrosis when the US cannot determine the level of the possible obstruction and there is no VUR on MCU then an i.v.u. is indicated. If this still fails to provide the required anatomical detail then antegrade studies will be required (fig. 5). In suspected recessive or dominant polycystic disease an i.v.u. is indicated in the late neonatal period.

Delayed images are mandatory even as late as 12–24 h after the administration of the contrast. Early films, in the presence of significant hydronephrosis, are frequently of little value.

Static Renal Isotope Scan

Tc99m DMSA is characterised in the newborn period by a high background activity, excess excretion of isotope in the urine and relatively low fixation of the isotope in the renal tubules. These physiological factors

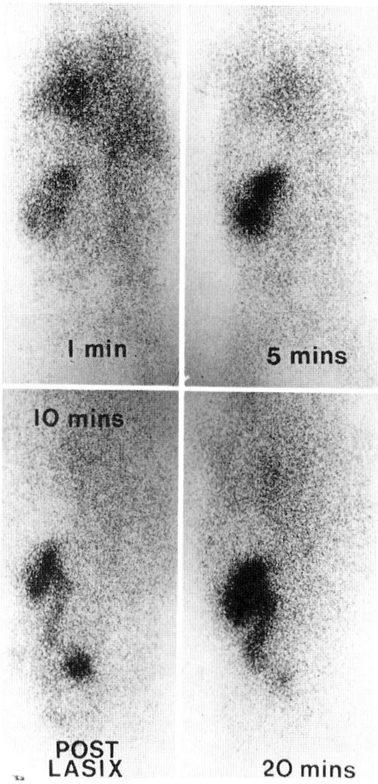

4a

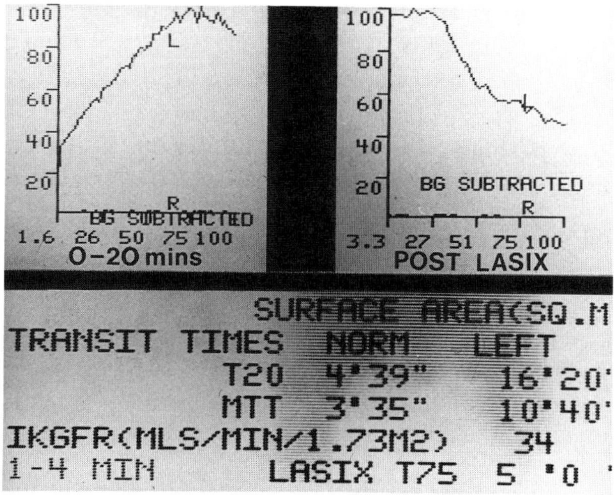

4b

Fig. 4a, b. This Tc99m DTPA was undertaken at 3 weeks of age in this male infant who was later found to have a non-functioning ectopic right kidney and a left kidney draining into a mega ureter. The DTPA scan shows good function with an individual kidney glomerular filtration rate of 34 ml/min per 1.73 m^2 and good wash out from the kidney following a diuretic stimulus. The ureter, however, does not wash out as well as the kidney, suggesting a hold up at the vesicoureteric junction.

Fig. 5. a. Bilateral hydronephrosis was diagnosed in utero, the post-natal US could not confirm dilated ureters behind the bladder but the gross bilateral hydronephrosis is clearly seen. *b* The radiograph 9 h after intravenous contrast fails to demonstrate the exact level of the obstruction on the left, on the right the kidney cannot be visualised. *c* The Tc99m DTPA scan shows bilateral poor function worse on the right than on the left with progressive accumulation of isotope in the renal pelvis with little isotope draining into the bladder. *d* The bilateral antegrade pyelogram study shows contrast outlining the dilated renal pelvis and the very narrow pelvi-ureteric junctions bilaterally. This child was suffering from bilateral pelvi-ureteric junction obstruction.

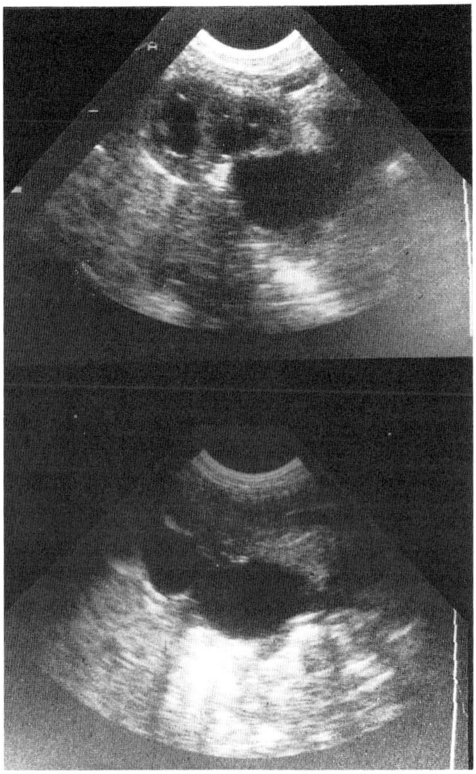

5a

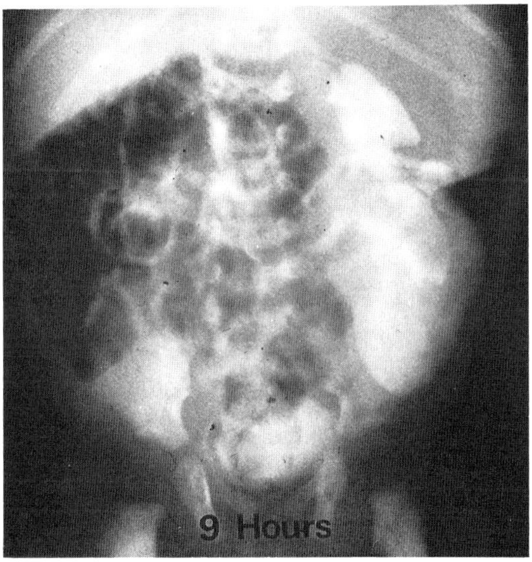

5b

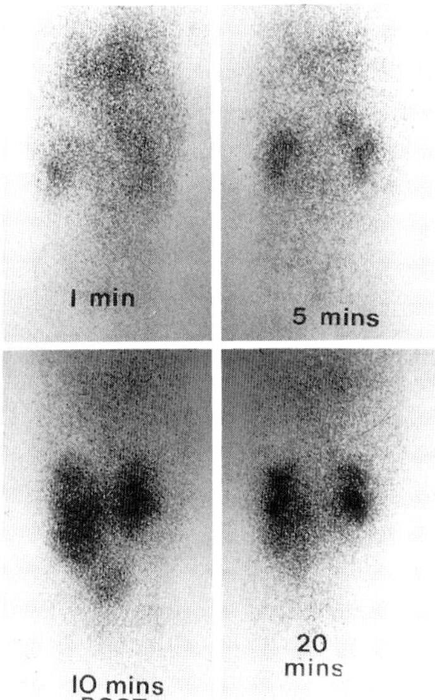

5c

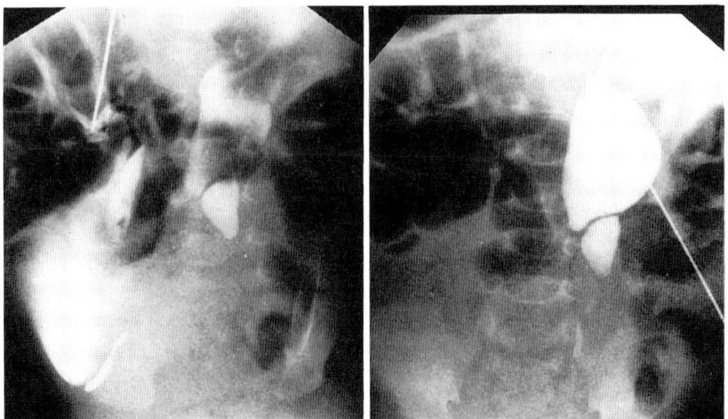

5d

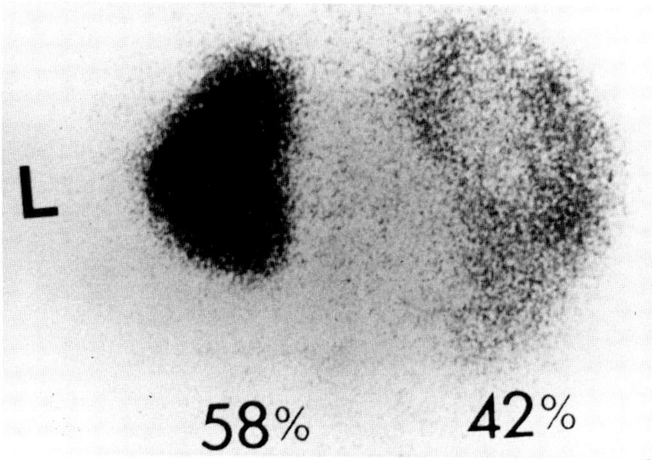

Fig. 6. This neonate had an antenatal diagnosis of a unilateral pelvi-ureteric junction obstruction. The Tc99m DMSA scan shows a normal left kidney, on the left there is decreased function in the kidney with the dilated renal pelvis seen as a negative defect. Note the high background and the relatively poor kidney-to-background ratio seen in the neonatal period.

prevent the high quality images seen over the ages of 18–24 months. This scan will still provide good differential function between the two kidneys (fig. 6) but will not positively exclude a renal scar during the first months of life.

Conclusion

The combined use of US and MCU allows confirmation of the antenatal diagnosis and any emergency therapy to be instituted when clinically indicated. Further imaging requires the use of radioisotopes to obtain information about individual kidney function and urinary drainage. This should be carried out as late as possible in the neonatal period. The i.v.u. is indicated only to provide specific answers and should never be undertaken in the early neonatal period.

Isky Gordon, MD, Department of Nephrology, Institute of Child Health, Hospital for Sick Children, Guilford Street, London WC1 1EH (UK)

Contr. Nephrol., vol. 67, pp. 183–187 (Karger, Basel 1988)

Multicystic Renal Dysplasia: Nephrectomy versus Conservative Treatment

Th. Lennert, M. Tetzner, M. Er, F. Jung, H.J. Kaufmann, W. Biewald
Children's Hospital, Free University, Berlin

Multicystic renal dysplasia is the most common cystic renal lesion in childhood. With rare exceptions it is a nonhereditary congenital malformation. Clinical reports in the literature show a prevalence of unilateral cases, whereas in autopsy material there is a higher prevalence of bilateral cases, which are by definition incompatible with life. Mir et al. [1] found 136 cases of renal cysts among more than 6,000 autopsies of children in Helsinki. Half of them, or 1% of the total material, proved to be renal dysplasia including 28 bilateral, 23 unilateral and 14 cases with focal distribution of dysplasia. The typical multicystic dysplastic kidney, also known as Potter's type II, is nonfunctioning by clinical and radiological means. Quite often the ureter is atretic. In about 30% the contralateral urinary tract shows various forms of malformation or vesico-ureteric reflux. There is also a frequent association with malformations of other organs, especially the heart, digestive tract or genital organs.

Histology shows cysts of various sizes, primitive epithelial ducts, islands of cartilage, clusters of immature glomeruli and tubules and regions of undifferentiated mesenchyme [2]. The differential diagnosis includes hydronephrosis, Wilms' tumor, renal vein thrombosis, neuroblastoma, polycystic kidneys and renal agenesis.

Due to frequent diagnostic difficulties to exclude hydronephrosis and tumors in former years, surgical exploration and subsequent nephrectomy has been the treatment of choice [3], although the benign character of the multicystic lesion has been established already for many years [4, 5]. The recent introduction of more sophisticated ultrasound technology has not only increased the number of cases with prenatally suspected cystic renal lesions, but has also broadened the diagnostic spectrum to identify these

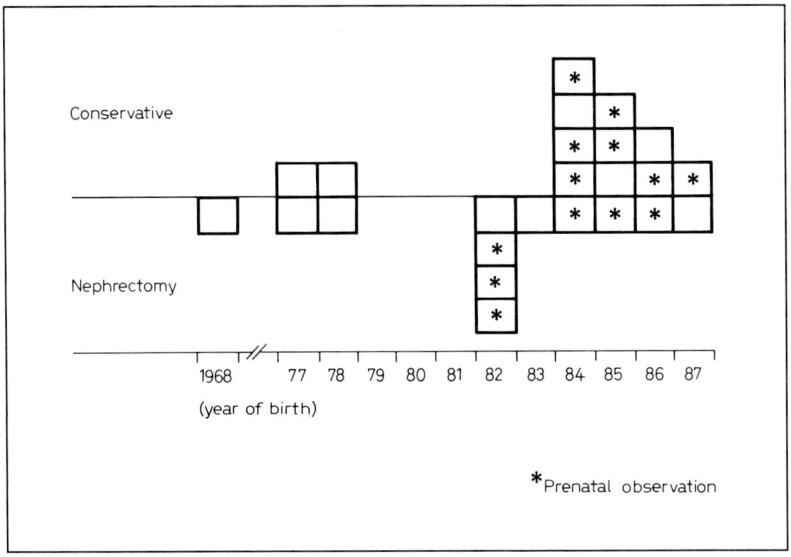

Fig. 1. Unilateral multicystic renal dysplasia. Children's Hospital, FU Berlin, 1968–1987.

malformations by imaging procedures without surgical intervention. Nevertheless, nephrectomy is still recommended in many recent publications like in the 1986 *Clinical paediatric nephrology* textbook edited by Postlethwaite [6].

At the Children's Hospital of the Free University of Berlin we have seen 24 cases of unilateral multicystic renal dysplasia in the last 20 years. The sex ratio was 16 boys to 8 girls. Five of them were born prematurely. The mean follow-up time was 32 months. Twelve of them were nephrectomized and 12 were treated conservatively. This decision was not always influenced by us. Figure 1 shows the distribution of cases over time. It is obvious that prenatal diagnosis has increased the number of cases who come under primary pediatric care immediately after birth.

Table I gives the details of the nephrectomy group. Half of the cases showed abnormal prenatal sonographic findings. In all cases the diagnosis was established within the first week of life and the children were operated within the first month. Eight cases had associated anomalies including 3 cases with reflux in the remaining kidney. In the conservative group there

Table I. Nephrectomy (n = 12)

Patient	Sex	Prenatal abnormality	Age, days		Associated anomalies
			at diagnosis	at operation	
1. C.D.	M	–	2	3	contralateral VUR II°
2. K.D.	F	–	6	24	hymenal atresia, uterus bipartit.
3. V.A.	M	–	1	8	mental retardation
4. O.S.	M	–	1	4	iris coloboma
5. T.V.	M	+	3	21	–
6. A.T.	F	+ (36 weeks)	6	16	–
7. N.R.	F	+ (38 weeks)	1	26	–
8. M.B.	F	–	1	6	contralateral dilation of calices
9. J.W.	F	+ (30 weeks)	1	1	contralateral VUR II°
10. M.R.	M	+ (28 weeks)	1	8	–
11. A.G.	M	+ (32 weeks)	1	25	Smith-Lemli-Opitz syndrome, horseshoe kidney
12. A.R.	M	–	3	14	contralateral VUR II°

were 7 with prenatal findings (table II). Within the first week of life the diagnosis was established in 9 of 12 cases, and at 4 months, 10 months and 7 years in the remaining 3 cases. Again there were associated anomalies in 7 of 12 cases, including 2 contralateral and 1 ipsilateral reflux. To compare the outcome of the two groups, we excluded all cases with a follow-up of less than 6 months. It is obvious (table III) that the remaining 8 cases in the nephrectomy group had a follow-up time nearly three times longer than the group of 9 conservative cases, as we changed our policy only recently. In regard to length and weight there was no difference between the two groups. Table IV shows no difference with regard to serum creatinine values and blood pressure at the time of the last examination. Compensatory renal hypertrophy of the contralateral kidney was seen in 7/8 nephrectomized and 7/9 conservative cases. Of the 3 cases without hypertrophy 2 had reflux and 1 showed a mild degree of dysplasia in the contralateral kidney. There were 3 episodes of urinary tract infection in 2 nephrectomized patients and 4 episodes in 3 patients of the conservative group. In regard to the difference in follow-up time there is a higher rate of infection

Table II. Conservative treatment (n = 12)

Patient	Sex	Prenatal abnormality	Age at diagnosis	Associated anomalies
1. J.B.	M	–	7 days	hexa/heptadactyly
2. S.S.	M	–	7 years	ipsilateral VUR I°
3. J.K.	F	+ (32 weeks)	1 day	–
4. M.R.	M	+ (29 weeks)	1 day	contralateral VUR II°
5. M.S.	M	–	10 months	–
6. L.Z.	M	+	2 days	–
7. D.S.	M	–	4 months	aortic stenosis
8. O.L.	M	+	1 day	thumb malformation, syndactyly
9. F.F.	M	+ (38 weeks)	1 day	contralateral VUR I°
10. L.D.	M	+ (28 weeks)	2 days	–
11. D.S.	F	+ (29 weeks)	1 day	–
12. B.G.	F	–	4 days	syndactyly, pulmonary stenosis PDA, VSD, microcephalus

Table III. Follow-up > 6 months

	Nephrectomy	Conservative
Patients, n	8	9
Mean follow-up (months)	67	25
Range	8–218	10–64
Length (mean percentile)	52	53
Weight (mean percentile)	59	58

Table IV. Follow-up > 6 months

	Nephrectomy	Conservative
Patients, n	8	9
Mean serum creatinine, μmol/l	72	60
Compensatory renal hypertrophy, n	7	7
Blood pressure > 90th percentile, n	1	0
UTI, episodes/patients	3/2	4/3

in the latter group. Only 1 patient, who belonged to the nephrectomy group, had a creatinine level above 120 μmol/l together with reflux and hypertension.

Conclusions

Today multicystic renal dysplasia can be diagnosed in most cases to a sufficient certainty by ultrasound. In doubtful cases additional information can be attained by i.v.p., MCU, scintigraphy and, occasionally, by cystoscopy. As Wilms' tumor and hydronephrosis can be ruled out by imaging techniques, there is no need for early surgical intervention, except for rare cases with mass problems. Preliminary follow-up data suggest that a unilateral multicystic kidney creates no real risk. The long-term renal function will depend solely on the functional quality of the contralateral kidney.

References

1 Mir, S.; Rapola, J.; Koskimies, O.: Renal cysts in paediatric autopsy material. Nephron *33:* 189–195 (1983).

2 Resnick, J.; Vernier, R.L.: Renal cystic diseases and renal dysplasia; in Holliday, Barratt, Vernier, Pediatric nephrology; 2nd ed., pp. 371–375 (Williams & Wilkins, Baltimore 1987).

3 Gipson, Th.G.; Anderson, E.E.; Bradford, W.D.: Multicystic renal dysplasia. Clin. Pediat. *15:* 896–900 (1976).

4 Bernstein, J.: The morphogenesis of renal parenchymal maldevelopment (renal dysplasia). Pediat. Clins N. Am. *18:* 395–407 (1971).

5 Kyaw, M.M.; Koehler, P.R.: Congenital multicystic kidney; in Gardner, Cystic diseases of the kidney, pp. 115–123 (Wiley, New York 1976).

6 Lendon, M.; Postlethwaite, R.J.: Cystic disorders of the kidneys; in Postlethwaite, Clinical paediatric nephrology, pp. 253–269 (Wright, Bristol 1986).

Th. Lennert, MD, Ahornstrasse 22, D–1000 Berlin 37

Contr. Nephrol., vol. 67, pp. 188–192 (Karger, Basel 1988)

Unilateral Multicystic Kidney Dysplasia: Follow-Up during the First Two Years of Life

H. Bachmann, J. Winkielman, H. Olbing
University Children's Hospital Essen, FRG

Unilateral abdominal tumors in neonates or infants are often of renal origin, caused by multicystic renal dysplasia (MRD) or hydronephrosis in most instances. Formerly, the multicystic kidney was removed immediately after diagnnosis, because by noninvasive diagnostic methods it was impossible to distinguish exactly between MRD and hydronephrosis [1–4, 9, 12]. Usually, surgical exploration was indicated and if MRD was found, nephrectomy was performed. Ultrasound helps to differentiate between the two conditions [11] and it is often possible to establish the diagnosis of MRD (or hydronephrosis) without surgical exploration and to decide whether removal of the multicystic kidney is necessary or not [3]. Information about the natural history of MRD is seen mostly as case reports [1, 3, 4, 6–8]. However, careful follow-up studies of patients with MRD, not nephrectomised, are necessary as a guideline for the appropriate decision.

At the end of 1982 we started a follow-up study in patients with unilateral MRD. Nephrectomy was limited to patients with unclear diagnosis and to patients with symptoms probably caused by MRD. Parenteral consent was obtained on entering the study.

We present our initial findings and the follow-up data of these patients from the neonatal period to the end of the second year of life.

Patients and Methods

From January 1983 to January 1986, unilateral MRD with normal contralateral kidney was diagnosed in 11 neonates or infants (1st to 6th weeks of life), 6 boys and 5 girls. Nephrectomy of the multicystic kidney was performed in 3 of the 11 patients, 2 of them since differentiation between MRD and hydronephrosis was not possible, and the third because of the large size of the multicystic kidney and recurrent vomiting. During the initial presentation all patients were examined by ultrasound, 10 by i.v.p., 7 by DMSA uptake and 9 by VCU. The 8 patients who were not nephrectomised were followed up regularly (at 3, 6, 12 and 24 months of age). Blood pressure, urinalysis and culture, serum creatinine and renal ultrasonography were performed at each follow-up visit.

Results

Initial Presentation (table I)

Diagnosis was suspected prenatally in 4 patients, during the 1 week of life in another 4, and between the 2nd and 6th weeks of life in 3. Nine of the 11 patients presented with a big palpable unilateral abdominal mass, 1 patient had urinary tract infection and another vomiting. The referring diagnosis was hydronephrosis in 5, cystic kidney in 3 and multicystic dysplasia in 3 patients.

Our diagnosis of MRD was based on a combination of the results of renal ultrasound, excretory urogram and radionuclide renal scan.

The most useful criteria for ultrasound identification of multicystic kidney was the presence of interfaces between the cysts, the nonmedial location of the largest cyst, the multiplicity of cysts, the absence of an identifiable renal sinus and the absence of parenchymal tissue [11]. The

Table I. Data at initial presentation

1	Patients		5	Indications for nephrectomy	
	Boys	6		Large abdominal mass with vomiting	1
	Girls	5		Atypical ultrasound findings	2
2	Time of diagnosis		6	Initial ultrasound findings	
	Prenatal	4		Kidney length	5.0–10.0 cm
	1st week of life	4		Kidney width	2.5–5.0 cm
	2nd to 6th week of life	3		Number of cysts	multiple
3	Symptoms at diagnosis			Cyst diameter	1.0–6.0 cm
	Unilateral abdominal mass	9		Solid structure < 10%	all
	Urinary tract infection	1	7	Diagnostic procedures	
	Vomiting	1		Ultrasound, typical findings	9/11
4	Initial diagnosis			i.v.p., no excretion of contrast material	10/10
	Hydronephrosis	5		VCU, VUR in the normal kidney	3/9
	Cystic kidney	3		DMSA uptake, no activity at the multicystic kidney	7/7
	Multicystic dysplasia	3			

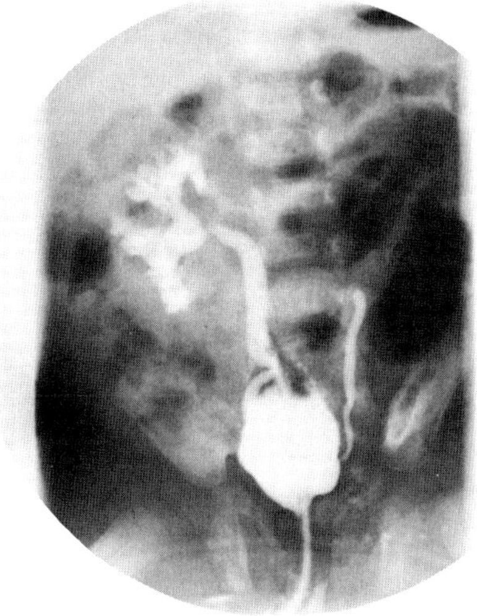

Fig. 1. VCU demonstrates reflux in the blind-ending ureter of the multicystic kidney (left) and vesicorenal reflux on the contralateral kidney (right).

length of the kidney was between 5.0 and 10.0 cm, the width between 2.5 and 5.0 cm and the diameter of the cyst up to 6.0 cm. The percentage of identifiable solid tissue was constantly below 10%. In 2 cases only one or two large cysts presented in the renal fossa. In these cases a questionable thin rim of parenchyma surrounded the cysts. Hydronephrosis could not be excluded clearly. Therefore, surgical exploration was performed, pathological examination of the removed kidney confirmed the diagnosis of MRD.

Urography showed that the involved side was nonfunctioning in all the cases. Radionuclide scan showed nonperfusion of the involved kidney with no activity of the isotope in the region of the renal fossa.

VCU demonstrated a blind-ending ureter in 2 of 9 patients on the side of the multicystic kidney and in 3 a vesicorenal reflux on the normal contralateral kidney (fig. 1).

Follow-Up (table II)

In 7 out of 8 patients, ultrasound showed no change in kidney length, kidney width, number of cysts, cyst diameter and percentage of solid structures in the multicystic kidney; in one patient the complete involution of the dysplastic kidney was observed. Two of the patients had urinary tract

Table II. Follow-up data

1 Ultrasound	2 Other findings	
No change in	Urine	
Kidney length	Proteinuria	none
Kidney width	Hematuria	none
Number of cysts	UTI	2 (1× with fever)
Cyst diameter	Blood pressure	normal
Percentage of solid structure	Serum creatinine	
(7 out of 8 patients)	At 12 months	normal for age
	At 24 months	normal for age

infection, 1 of them with fever; this patient had a vesicorenal reflux at the contralateral kidney. Blood pressure remained normal for age, and serum creatinine at 12 and 24 months of age was within the normal range as for children with solitary kidneys.

Discussion

We and others have shown that diagnosis of MRD can be done accurately by combination of the results of ultrasound, excretory urogram and radionuclide isotope scan [3, 11]. Our data support the opinion that the separation of MRD from solid and potentially malignant renal masses seem to be more simple than the differentiation from hydronephrosis [3, 6, 9]. The absence of solid structures is one of the main criteria of MRD. Therefore, surgical exploration should be done in any case in which solid tissue can be demonstrated in a higher percentage. If only one or two large cysts are in the renal fossa differentiation of hydronephrosis and MRD may be impossible. Certainly, when the diagnosis of MRD is in doubt, there is no controversy about the need for surgical exploration and removal of the involved kidney [6, 10]. In our experience ultrasound is the most important simple imaging method for the diagnosis of MRD. Excretory urogram is of value especially in assessing the status of the uninvolved kidney. Further studies are needed to show if renal scanning by isotope studies is mandatory for the diagnosis of MRD. VCU seems to be important for evaluating the risk of urinary tract infection in such patients with a functionally solitary kidney, 3 of 9 patients investigated by VCU had vesicorenal reflux at the normal contralateral kidney and 1 of them developed UTI with fever.

During the follow-up period during the first and second years of life, our patients had no identifiable disadvantage from the multicystic kidney and in 7 out of 8 patients there were no ultrasonographic changes in size of the multicystic kidney or of the cysts. But the number of patients we observed is small and the follow-up period relatively short. Long-term follow-up studies should be instituted to inform about late sequelae in patients not nephrectomised with MRD [10].

The management of MRD in infancy and childhood should be further investigated. In the meantime, any patient in which the diagnosis of MRD is in doubt or if the patient is symptomatic because of MRD, nephrectomy should be performed. In patients without problems immediate or early nephrectomy is not mandatory, if necessary it can be done electively at a later date.

References

1 Ambrose, S.S.; Fould, R.A.; Trubock, T.S.; Parrott, T.S.: Unilateral multicystic renal disease in adults. J. Urol. *128:* 366 (1982).

2 Bernstein, J.: Renal hypoplasia and dysplasia; in Edelmann, Pediatric kidney disease, vol. II, pp. 541–557 (Little, Brown, Boston 1978).

3 Bloom, D.A.; Brosman, S.: The multicystic kidney. J. Urol. *120:* 211 (1978).

4 Gipson, T.G.; Anderson, E.E.; Bradford, W.D.: Multicystic renal dysplasia. Pathologic and clinical observations in 22 cases. Clin. Pediat. *15:* 896 (1976).

5 Greene, L.B.; Feinzaig, W.; Dahlin, L.D.: Multicystic dysplasia of the kidney. With special reference to the contralateral kidney. J. Urol. *105:* 482 (1971).

6 Hartmann, G.E.; Smolik, L.M.; Shochat, S.J.: The dilemma of the multicystic dysplasic kidney. Am. J. Dis. Child. *140:* 925 (1985).

7 Javadpour, N.; Chelouhy, E.; Moncada, L.; Rosenthal, I.M.; Bush, I.M.: Hypertension in a child caused by a multicystic kidney. J. Urol. *104:* 918 (1970).

8 Scholtmeijer, R.J.; Harten, J.J. van der: Unilateral multicystic kidney and contralateral hydronephrosis in the newborn. Br. J. Urol. *47:* 176 (1975).

9 Spence, H.M.: Congenital unilateral multicystic kidney: an entity to be distinguished from polycystic kidney disease and other cystic disorders. J. Urol. *74:* 693 (1955).

10 Stanisic, T.H.: Review of 'The dilemma of the multicystic dysplasic kidney' (Editorial). Am. J. Dis. Child. *140:* 865 (1985).

11 Stuck, K.J.; Koff, S.A.; Silver, T.M.: Ultrasonic features of multicystic dysplasia kidney. Expanded diagnostic criteria. Radiology *143:* 217 (1982).

12 Walker, D.; Fennell, R.; Garin, E.; Richard, G.: Spectrum of multicystic renal dysplasia. Diagnosis and management. Urology *11:* 433 (1978).

Prof. H. Bachmann, MD, Kinderklinik, Zentralkrankenhaus 'Links der Weser', Senator-Wessling-Strasse 1, D–2800 Bremen 61 (FRG)

Contr. Nephrol., vol. 67, pp. 193–199 (Karger, Basel 1988)

Studies on Fetal Renal Function

Comparison of Microproteins in Amniotic Fluid and Fetal Urine

R. Burghard, J.U. Leititis, M. Brandis

Department of Pediatrics, Philipps-University, Marburg/Lahn, FRG

The increasing utilization of sonography during pregnancy has led to a large number of reports on prenatal diagnosis of fetal obstructive uropathies and antenatal detection has become a relatively common mode of presentation of congenital renal abnormalities [4]. Some degree of uncertainty, however, exists concerning a rational approach to the management of fetal urinary tract obstruction. The technology of fetal surgery and in utero percutaneous catheter drainage is already well established [6], but serious limitations exist in the identification of the fetus with anomalies appropriate for in utero intervention.

Hitherto, there is no definitive method to identify the fetus with potentially salvageable renal function that is in danger of deterioration with further advancing pregnancy and there is no objective evidence that a third-trimester intervention will have a more favorable effect on renal function, as compared with that which can be obtained by early neonatal corrective surgery [4]. Thus, the major problems that remain are objective means of evaluating fetal renal function and a way of selecting the fetus who may benefit from either in utero surgery or delivery before term.

It has been suggested that some biochemical characteristics of amniotic fluid (AF) may be influenced by developmental changes of fetal renal function and maturation [8, 9]. Evidence has accumulated from a limited number of investigations that some proteins of low molecular weight are largely produced and excreted by the fetus itself and that their concentrations in AF may reflect indirectly the state of functional development of the fetal kidneys [5, 7]. This reasoning led us to study the protein content and protein composition of AF in normal pregnancies and pregnancies complicated by fetal urinary tract abnormalities.

Material and Methods

Amniotic Fluid and Fetal Urine

AF was obtained from 171 healthy pregnant women between the 16th and 38th weeks of gestation by abdominal amniocentesis and vaginally during normal delivery in some cases of advanced pregnancy (n = 7). Amniocentesis was mainly performed for genetic reasons (n = 142), but also for evaluation of fetal pulmonary maturation (n = 22). The duration of pregnancy was determined on the basis of menstrual history and sonographic evaluation. Patients were excluded from this study if menstrual data appeared to be unreliable or if the gestational age calculated by history and sonography differed by more than 0.5 weeks. Samples contaminated by blood or meconium were also discarded.

Additionally, 11 samples of AF and fetal urine were obtained simultaneously of 9 pregnancies complicated by bilateral fetal hydronephrosis. The diagnosis of fetal urinary tract obstruction was confirmed by ultrasonography between the 16th and 33rd weeks of gestation on repeated examination.

Analytical Procedures

Creatinine concentrations were determined by Jaffé picric acid reaction. Total protein concentration (TP) was evaluated by a protein-dye-binding reaction according to the method of Bradford [1]. Alpha-1-microglobulin (α-1-MG) and beta-2-microglobulin (β-2-MG) were assayed by single radial immunodiffusion and an enzyme immunoassay, respectively (Behringwerke AG, Marburg/Lahn, FRG).

Differentiation of proteins was performed in 1-µl samples of unconcentrated AF and fetal urine, using microgradient gel electrophoresis (MGGE) with a continuous concentration gradient of 4–40% polyacrylamide according to the method described by Neuhoff [10]. Gels were scanned with a Joyce-Loebl double-beam microdensitometer and the area of protein peaks was determined by simultaneous automatic integration. By means of MGGE, proteins were separated into three main fractions: albumin (molecular weight 68 kdaltons), proteins of higher molecular weight than albumin (HMW proteins), and proteins with a lower molecular weight (LMW proteins) than albumin. For quantitative evaluation of these three constituents, protein peaks were compared with an albumin standard. The standard curve exhibited an excellent linear correlation between albumin concentrations in the range 0.25–1.0 g/l and the integrated peak area ($r = 0.98$, $p < 0.001$). LMW and HMW protein concentration values are approximate because no specific standards are available.

Statistical Calculation

All values were calculated as median because of the non-normal distribution. A nonparametric test was used to detect differences (Kruskal-Wallis test). In the correlation analysis a curve-fitting programme was used, calculating linear, logarithmic or exponential functions.

Table I. Concentrations of creatinine and proteins in amniotic fluid during the 2nd and 3rd trimesters of gestation (median, range; p calculated by Kruskal-Wallis test)

	Weeks of gestation							p
	16–17	18–19	20–23	24–27	28–32	33–36	37–38	
Creatinine, mg/l	6.9 (3.3–9.5)	6.9 (5.2–12.0)	7.6 (5.1–8.5)	8.0 (7.1–9.0)	11.4 (4.1–14.5)	13.1 (5.1–22.8)	15.9 (12.8–19.9)	< 0.001
Total protein, g/l	2.35 (0.35–5.96)	2.92 (0.90–6.86)	3.40 (2.25–6.67)	2.78 (1.45–5.46)	1.30 (0.85–4.31)	1.78 (0.90–3.84)	2.20 (1.15–2.45)	< 0.001
Albumin, g/l	1.39 (0.60–4.07)	1.57 (0.46–4.09)	2.60 (0.80–6.82)	1.29 (0.62–2.98)	0.81 (0.46–2.62)	0.89 (0.41–2.21)	1.03 (0.15–1.29)	< 0.001
HMW proteins g/l	0.17 (0.01–0.55)	0.18 (0.06–0.43)	0.21 (0.08–0.97)	0.17 (0.08–0.60)	0.20 (0.08–0.96)	0.20 (0.05–0.81)	0.17 (0.05–0.58)	NS
LMW proteins g/l	0.18 (0.06–0.63)	0.18 (0.01–0.79)	0.22 (0.09–0.78)	0.15 (0.02–0.21)	0.13 (0.02–0.21)	0.12 (0.03–0.21)	0.03 (0.01–0.16)	< 0.0025
α-1-MG, mg/l	34.5 (16.1–110.3)	30.1 (16.9–71.4)	28.4 (20.5–79.4)	30.7 (21.6–58.3)	21.6 (15.9–27.4)	16.9 (12.5–24.3)	14.1 (10.8–18.7)	< 0.001
β-2-MG, mg/l	7.9 (1.0–24.7)	7.2 (1.0–15.7)	7.8 (1.0–16.9)	11.6 (3.4–14.1)	5.5 (3.5–7.6)	4.5 (1.0–7.3)	2.4 (1.0–6.1)	< 0.001

Results

In table I, normal values for creatinine and protein concentrations in AF are given in relation to gestational age. It is evident that creatinine concentrations remained fairly constant up to the 28th week of gestation and showed a progressive and exponential increase during the last trimester of gestation. Concentrations of TP, albumin, LMW proteins, α-1-MG and β-2-MG decreased significantly at the end of the second trimester.

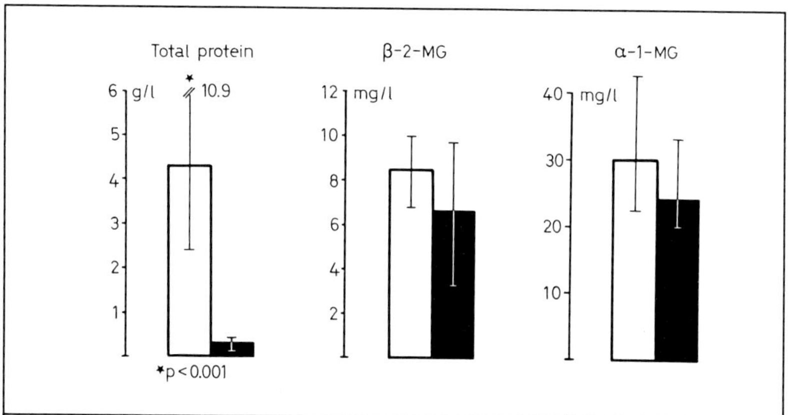

1

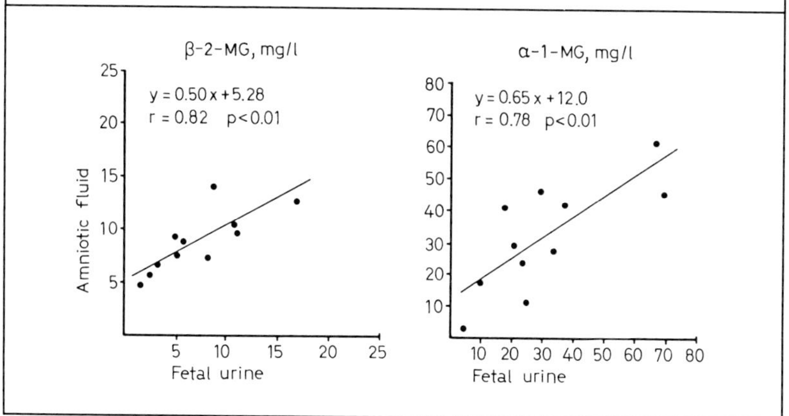

2

Fig. 1. Concentrations of total protein (TP), β-2-MG and α-1-MG in amniotic fluid (□) and fetal urine (■). Median, 95% confidence limits.

Fig. 2. Correlations between the concentrations of micromolecular proteins in amniotic fluid and fetal urine.

There was no further decline in TP and albumin concentrations after the 32nd week of gestation, whereas LMW proteins, including α-1-MG and β-2-MG, decreased gradually as pregnancy advanced. HMW protein concentrations were unchanged throughout the last two trimesters of gestation.

Of all protein fractions analysed in AF, a good correlation with TP concentrations existed only for albumin ($r = 0.68$, $p < 0.001$), indicating that changes in TP content of AF during pregnancy are mainly due to albumin. The LMW protein fraction in AF as characterized by MGGE was positively correlated with α-1-MG ($r = 0.74$, $p < 0.001$) and β-2-MG ($r = 0.72$, $p < 0.001$), which can therefore be used as micromolecular marker proteins representing the LMW protein content of AF.

If protein concentrations in AF and the urine of fetuses with obstructive uropathies are compared (fig. 1), TP- and albumin concentrations in AF exhibited significantly higher values, whereas α-1-MG and β-2-MG were not different in AF and fetal urine. Concentrations of micromolecular proteins in AF were significantly correlated with the concentrations in fetal urine (fig. 2), but this was not the case if the creatinine or protein/creatinine ratios in AF and fetal urine were compared.

The concentrations of α-1-MG and β-2-MG in AF of fetuses with urinary tract obstruction were in the normal range, except in 1 case (fig. 3). Postnatal renal function studies were performed in 5 cases and revealed a normal kidney function immediately after birth or following early surgical treatment.

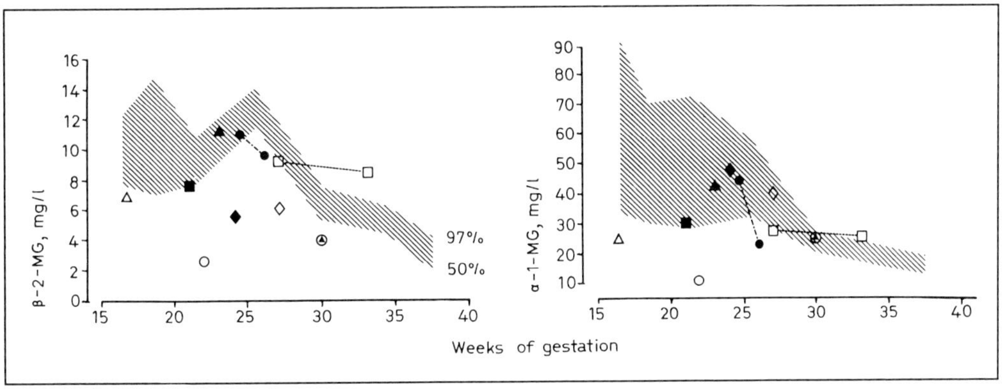

Fig. 3. Concentrations of α-1-MG and β-2-MG in amniotic fluid of 9 fetuses with obstructive uropathies. ▧ = Upper normal limit.

Discussion

In a previous investigation [2] concentrations of micromolecular proteins were found to be the same in AF and in the first postnatal urine of newborns of corresponding gestational age. Under the implicit assumption that immediate neonatal renal function is similar to that of the fetus, we concluded from these data that LMW proteins in general derive from the fetus itself and that fetal urinary production contributes considerably to the LMW protein content of AF.

In comparison with the protein composition of AF, after-birth analysis of micromolecular proteins in the urine of infants delivered prematurely and at term yielded an identical developmental pattern. These data, when considered in view of the evidence that renal tubular metabolism is the major catabolic pathway of these proteins, suggest that maturation of this pathway is the most likely explanation for the decrease of the LMW protein concentration in AF during the last trimester of gestation [2]. LMW proteins seem to accumulate in AF during development until reabsorption ability increases in the course of the morphological and functional maturation of proximal tubular cells.

In the present investigation, a strong correlation existed between the concentrations of α-1-MG and β-2-MG in AF and fetal urine confirming definitively the hypothesis that fetal renal function is the major determinant for the LMW protein content of AF. Analysis of micromolecular proteins in AF, therefore, should theoretically give further information on fetal kidney function and dysfunction. From our results it is undecided, however, whether distinct pathological conditions affecting the fetal kidneys, i.e. in obstructive uropathies, may result in specific changes of AF micromolecular protein concentration indicating delayed or defective intrauterine renal maturation. Further investigation is required to define exactly the diagnostic value of AF protein analysis in evaluating fetal renal function.

References

1 Bradford, M.M.: A rapid and sensitive method for quantitation of microgram quantities of protein utilizing the principle of protein-dye binding. Analyt. Biochem. *72:* 248–254 (1976).

2 Burghard, R.; Pallacks, R.; Gordjani, N.; Leititis, J.U.; Hackelöer, B.J.; Brandis, M.: Microproteins in amniotic fluid as an index of changes in fetal renal function during development. Pediat. Nephrol. (in press, 1987).

3 Cejka, J.; Kithier, K.; Belaniaric, J.; Al-Sarraf, M.: Fetospecific features of human beta-2-microglobulin. Experientia *30:* 458–460 (1974).

4 Diament, M.J.; Fine, R.N.; Ehrlich, R.; Kangarloo, H.: Fetal hydronephrosis: Problems in diagnosis and management. J. Pediat. *103:* 435–440 (1983).

5 Hall, P.W.; Roux, J.F.: Amniotic fluid beta-2-microglobulin concentration. An index of gestational age. Am. J. Obstet. Gynec. *120:* 56–59 (1974).

6 Harrison, M.R.; Golbus, M.S.; Filly, R.A.; Callen, P.W.; Katz, M.; De Lorimer, A.A.; Rosen, M.; Jonsen, A.R.: Fetal surgery for congenital hydronephrosis. New Engl. J. Med. *306:* 591–593 (1982).

7 Jonasson, L.E.; Evrin, P.E.; Wibell, L.: Content of β-2-microglobulin and albumin in human amniotic fluid. Acta obstet. gynaec. scand. *53:* 49–58 (1974).

8 Lind, T.; Parkin, F.M.; Cheyne, G.A.: Biochemical and cytological changes in liquor amnii with advancing gestation. J. Obstet. Gynaec. Br. Commonw. *76:* 673–683 (1969).

9 Lind, T.; Kendall, A.; Hytten, F.E.: The role of the fetus in the formation of amniotic fluid. J. Obstet. Gynaec. Br. Commonw. *79:* 289–298 (1972).

10 Neuhoff, V.: Microelectrophoresis on gradient gels; in Neuhoff, Micromethods in molecular biology, pp. 1–83 (Springer, Berlin 1973).

R. Burghard, MD, Zentrum für Kinderheilkunde der Philipps-Universität, Deutschhausstrasse 12, D–3550 Marburg/Lahn (FRG)

Contr. Nephrol., vol. 67, pp. 200–205 (Karger, Basel 1988)

Alport's Syndrome: Localization of the X-Chromosomal Gene and Consequences for Future Investigations

H. Brunner[a], *C. Schröder*[b], *L. Monnens*[b], *J. Veerkamp*[c], *H.-H. Ropers*[a]

Departments of [a]Human Genetics, [b]Pediatrics, and [c]Biochemistry, St. Radboud Hospital, University of Nijmegen, The Netherlands

Alport's syndrome is a genetic disorder of the kidney with many poorly understood features. Especially, the pathogenesis and the exact genetic pattern of the syndrome are unclear. Recent studies have contributed new data with respect to these features. The syndrome is correlated with an abnormal composition of the so-called globular (NC1) domain of the collagen type IV molecule present in the glomerular basement membrane, which indicates a metabolic disturbance of the basement membrane [9, 14].

Three modes of inheritance are assumed to exist in Alport's syndrome [4, 11]: an X-linked-dominant, an autosomal-dominant, and an autosomal-recessive mode. Of these three, autosomal-recessive inheritance is probably rare. The relative contributions of X-linked-dominant and autosomal-dominant inheritance to Alport's syndrome are unknown, although some authors consider X-linked inheritance as the most common [3].

Recently, Menlove et al. [12] presented preliminary results on linkage studies of X-linked Alport's syndrome to DNA probes from the centromeric region of the X-chromosome in three Utah kindreds. In the present study, data from three European families are presented.

Patients and Methods

Patients

Fifty-four persons out of three families with Alport's syndrome were investigated. Diagnosis was confirmed by renal biopsy in at least one member of each family. The typical ultrastructural lesions, i.e. focal thickening and splitting of the glomerular basement membrane as well as intramembranous granules, were found in each biopsy specimen. Family members were considered affected if a freshly voided urinary specimen contained more than 10 erythrocytes per mm^3 [5]. Renal function was determined by creatinine clearance calculated on 24-hour urinary specimens. Only in a few patients were single serum creatinine measurements obtained. Audiographic examination was performed routinely in all the investigated patients. Ophthalmologic examination was performed only if symptoms were present.

Methods

DNA was extracted from leukocytes of 10 ml of EDTA-anticoagulated venous blood by standard procedures [1]. Four micrograms of genomic DNA was digested to completion with restriction enzymes Taq I or Msp I (Boehringer, Mannheim, FRG) under conditions specified by the manufacturer. DNA was electrophoresed in 0.7% agarose gels and transferred to Gene Screen (New England Nuclear) membranes by Southern blotting [15]. ^{32}P-labeled DNA probes were hybridized for 24 h at 65 °C [17]. Autoradiography was performed for 3 days at −70 °C using double intensifying screens. For linkage analysis a recent version (3.5) of the linkage program package was employed, which was kindly provided by Dr. Jürg Ott (Zürich, Switzerland).

Results

The participating persons were tested employing 8 X-chromosomal markers. Four of these markers were informative, defining the DXS 7, DXYS 1, DXS 17 and DXS 11 loci, respectively. Tight linkage was found between the gene coding for X-linked Alport's syndrome (XAS) and the DXS 17 locus, defined by probe S9 which has been assigned to bands q 21.3–q 22 of the X-chromosome [2, 17].

One cross-over was observed in 26 informative meioses giving a maximum lod (log of odds) score of 3.63 at a recombination fraction of 0.04 (95% confidence limits 0.002–0.19). Thus, the relative odds for linkage to this site are over 4,000:1. A weakly positive lod score (1.50) was found for the DXS 11 locus, defined by probe 22.33. The other markers showed no linkage to the gene for XAS.

Discussion

From our data we conclude that the XAS gene is localized in the middle portion of the X-chromosome long arm in the vicinity of DXS 17 (bands q 21.3–q 22) and DXS 11 (Xq 24–26) (fig. 1). Comparison of our data with the preliminary data presented by Menlove et al. [12] suggests

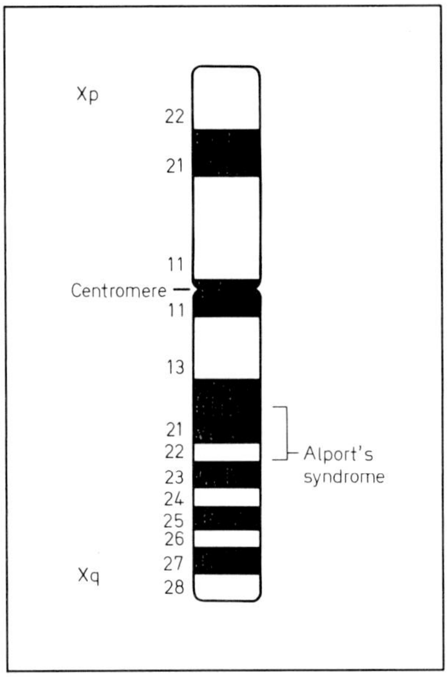

Fig. 1. Proposed localization of the gene of X-linked Alport's syndrome.

genetic homogeneity, since both sets of results are compatible with a localization of the Alport's syndrome gene in Xq 21.3–Xq 22. X-linked Alport's syndrome has been subdivided into several clinical subtypes by Hasstedt et al. [8] on the basis of the presence or absence of deafness and age of end-stage renal disease. Comparing families with different clinical features will eventually determine whether this apparent heterogeneity exists at the molecular level, and, if so, whether it is due to separate genes or to different mutations within a single gene.

Our data suggest that DNA analysis may distinguish between X-linked and autosomal-dominant inheritance in a given pedigree, and therefore should enhance accurate counselling in such families. Furthermore, DNA analysis should prove to be of value in carrier detection and prenatal diagnosis of Alport's syndrome in families in which the disease is transmitted in an X-linked fashion. Clearly, additional families should be studied to reduce confidence intervals.

The search for the Alport gene proper can now be initiated in several ways: once flanking markers have been obtained the gene may be approached through molecular genetic techniques, such as gene walking, pulsed field gel electrophoresis and by a search for so-called HTF islands (regions of high CG content) that are commonly associated with genes. Any candidate gene can then be tested for expression in kidney cells and for differences between patients and unaffected individuals. Although straightforward, this approach is time consuming as a large number of candidate genes may have to be analyzed.

Biochemical investigations of the glomerular basement membrane indicate that the basic pathogenetic defect in Alport's syndrome is located in the noncollagenous globular domain (NC1) of type IV collagen [9, 14]. Analysis of genes important in collagen metabolism may exclude such a gene as the cause of Alport's syndrome if it is not on the relevant X-chromosomal segment. In this manner, DNA analysis has excluded a mutation in the α_1 (IV) collagen gene as the cause of Alport's syndrome, since this gene has been located on chromosome 13 [13]. The localization of the α_2 (IV) collagen gene is at present unknown, although it has been suggested that it may be either on chromosome 8 or X [13, 16]. It will be interesting to see if one of these localizations is correct. Recently, it has been proposed that collagen type IV not only consists of these two chains, but that a third and a fourth chain exist [10]. These chains could also be evaluated as candidates for the localization of the biochemical defect.

Monoclonal antibodies against parts of the globular domain of the collagen IV molecule have been developed that do not react with the glomerular basement membrane of Alport's syndrome patients [9, 14], although this may not be true in all patients [7]. Once specificity has been established, such antibodies can be used to screen cDNA expression libraries that consist of colonies of recombinant bacteria, each bacteria containing a small piece of human cDNA. By identifying colonies that react with the antibody, the responsible gene or a portion of it may be isolated. This DNA can then be analyzed to verify whether it originates from the relevant segment of the X-chromosome. Although conceptually attractive, this approach is likely to fail if the basic defect in Alport's syndrome is not in a gene coding for a structural protein, but rather in a gene coding for a modifying enzyme. It may be worthwhile to look for collagen-modifying enzymes encoded by the X-chromosome. However, deficiency of lysyl oxidase, the only such enzyme for which the gene is currently believed to be on the X-chromosome, does not cause prominent renal symptoms and thus is not a likely candidate for Alport's syndrome [6]. Collagenous digests of the glomerular basement membrane of Alport's syndrome patients do not show at SDS-polyacrylamide electrophoresis and immunoblotting with Goodpasture's antiserum a typical 28-kdalton band which is present in normal individuals. Full or partial deletion of a modifying enzyme can be hypothesized as a possible cause for this phenomenon [9].

At present, the possibility that the biochemical defect in Alport's syndrome is in a noncollagenous protein (possibly associated with collagen type IV) should not be overlooked.

In view of the complex problems facing research into the genetic defect in Alport's syndrome, a multidisciplinary approach combining modern genetic techniques, protein biochemistry and immunological methods is advocated and should lead to substantial progress in the near future.

Note Added in Proof

The gene coding for the α_2 (IV) collagen chain has now been assigned to chromosome 13 [18].

Acknowledgements

Dr. D. Menzel and Prof. Dr. H. Olbing (Department of Pediatric Nephrology, University of Essen, FRG) are kindly acknowledged for clinical evaluation of one of the families, J. Tuerlings and C. van Bennekom for expert technical assistance, H. Lambermon for help with linkage calculations and B. Wieringa for many helpful discussions.

References

1 Aldridge, J.; Kunkel, L.; Bruns, G.; et al.: A strategy to reveal high-frequency RFLP's along the human X-chromosome. Am. J. hum. Genet. *36:* 546–564 (1984).

2 Drayna, D.; Davies, K.; Hartley, D.; et al.: Genetic mapping of the X-chromosome by using restriction fragment length polymorphisms. Proc. natn. Acad. Sci. USA *81:* 2836–2839 (1984).

3 Feingold, J.; Bois, E.; Chompret, A.; Broyer, M.; Gubler, M.C.; Grünfeld, J.P.: Genetic heterogeneity of Alport syndrome. Kidney int. *27:* 672–677 (1985).

4 Feingold, J.; Bois, E.: Genetics of Alport's syndrome. Pediat. Nephrol. *1:* 436–438 (1987).

5 Gadeholt, H.: Quantitative estimation of cells in urine. Acta med. scand. *183:* 369–374 (1968).

6 Goodfellow, P.N.; Davies, K.E.; Ropers, H.-H.: Report of the committee on the genetic constitution of the X and Y chromosomes. 8th Int. Workshop on Human Gene Mapping. Cytogenet. Cell Genet. *40:* 314 (1985).

7 Grünfeld, J.P.; Grateau, G.; Noël, L.H.; et al.: Variants of Alport's syndrome. Pediat. Nephrol. *1:* 419–421 (1987).

8 Hasstedt, S.J.; Atkin, C.L.; San Juan, A.C., Jr.: Genetic heterogeneity among kindreds with Alport syndrome. Am. J. hum. Genet. *38:* 940–953 (1986).

9 Kleppel, M.M.; Kashtan, C.E.; Butkowski, R.J.; Fish, A.J.; Michael, A.F.: Alport familial nephritis. Absence of 28 kilodalton non-collagenous monomers of type IV collagen in glomerular basement membrane. J. clin. Invest. *80:* 263–266 (1987).

10 Langeveld, J.P.M.; Yem, A.W.T.; Timoneda, J.; Saus, J.; Wisdom, B.J.; Hudson, B.G.: Molecular structure of collagen IV of glomerular basement membrane; in Price, Hudson, Renal basement membranes in health and disease, pp. 25–39 (Academic Press, London 1987).

11 McKusick, V.A.: Mendelian inheritance in man; 7th ed. (Johns Hopkins University Press, Baltimore 1986).

12 Menlove, L.; Kirschner, N.; Nguyen, K.; et al.: Linkage between Alport syndrome-like hereditary nephritis and X-linked RFLP's. Cytogenet. Cell Genet. *40:* 697–698A (1985).

13 Pihlajaniemi, T.; Tryggvason, K.; Myers, J.C.; et al.: cDNA clones coding for the pro-α1(IV) chain of human type IV procollagen reveal an unusual homology of amino acid sequences in two halves of the carboxyl-terminal domain. J. biol. Chem. *260:* 7681–7687 (1985).

14 Savage, C.O.S.; Pusey, C.D.; Kershaw, M.J.; et al.: The Goodpasture antigen in Alport's syndrome: Studies with a monoclonal antibody. Kidney int. *30:* 107–113 (1986).

15 Southern, E.: Detecting of specific sequences among DNA fragments separated by gel electrophoresis. J. molec. Biol. *98:* 503–517 (1975).

16 Uitto, J.: Biochemistry of collagen in diseases. Ann. intern. Med. *105:* 740–756 (1986).

17 Wieacker, P.; Davies, K.; Cooke, H.J.; et al.: Towards a linkage map of the human X-chromosome. Regional assessment of 16 cloned single copy DNA sequences employing a panel of somatic cell hybrids. Am. J. hum. Genet. *36:* 265–276 (1984).

18 Killen, P.D.; Francomano, C.A.; Yamada, Y.; Modi, W.S.; O'Brien, S.J.: Partial structure of the human α_2 (IV) collagen chain and chromosomal localization of the gene (col yA2). Hum. Genet. *77:* 318–324 (1987).

H. Brunner, MD, Department of Human Genetics, St. Radboud Hospital,
University of Nijmegen, POB 9101, NL–6500 HB Nijmegen (The Netherlands)

Follow-Up Studies

Contr. Nephrol., vol. 67, pp. 206–218 (Karger, Basel 1988)

Multicenter Cooperative Studies in Pediatric Nephrology

Johannes Brodehl

Department of Pediatric Nephrology and Metabolic Disorders, Kinderklinik, Medizinische Hochschule Hannover, Hannover, FRG

There are mainly two instruments which can be used in clinical medicine to evaluate the efficacy and the risk of a certain treatment by scientific means. These are: (1) the close observation of the individual patient, which, however, is always anecdotal and uncontrolled, and (2) the controlled clinical trial, which compares randomized groups of patients prospectively.

Both methods are widely used in pediatric nephrology and have certainly increased both the quality and quantity of the body of our knowledge in this field [11]. In the following I will focus on the practical aspects of cooperative controlled clinical trials and first present the results of a questionnaire sent out to all members of the European Society for Paediatric Nephrology (ESPN) who are concerned with cooperative studies in the Society. Then I will touch on some essentials which seem to be mandatory in order to start and run a prospective clinical trial. I think there is much evidence that such trials are most powerful tools in the evaluation of clinical procedures and treatments in case they are planned and conducted properly.

Cooperative Studies of ESPN Members

A questionnaire was sent out to all registered members of ESPN early in 1987. Of 199 members, the questionnaire was answered by 101 (51%). About two-thirds of those who answered (67/101) stated that they had been involved in cooperative multicenter studies during the indicated period of

Table I. ESPN members involved in surveys on epidemiology and natural history (from 101 answers)

Topic	Number of members
Chronic renal failure	11
Hemolytic-uremic syndrome	10
Nephrotic syndrome	9
Hematuria	7
Membranoproliferative glomerulonephritis	6
Vesicoureteral reflux	4
Nephritis	3
Proteinuria	1
Total	51

10 years, i.e. from 1977 to 1986. This demonstrates the importance and wide usage of cooperative multicenter studies as a tool in scientific pediatric nephrology.

Multicenter studies were used in almost all countries of Europe in which active groups of pediatric nephrologists were working. From the answers on the questionnaire it became apparent that the activity of members in regard to cooperative studies seems to depend somewhat on the pool size of pediatric nephrologists in the different countries, i.e. in those countries with a larger group of pediatric nephrologists, the percentage of participating members is higher than in those with a smaller group.

In the questionnaire the types of cooperative multicenter activities were not specified. Two types of studies emerged from the answers: firstly, the cooperative surveys (retrospective or prospective) on the epidemiology and natural history of certain kidney diseases, in order to find out about prevalence and incidence, and, secondly, the interventional therapeutic trials, which were either uncontrolled or controlled by a comparison group.

Fifty-one members of the ESPN were involved in multicenter studies of the first type. i.e. in surveys. These are listed in table I. They were mainly concerned with acute and chronic renal failure [7, 8, 17, 35], the hemolytic uremic syndrome and the nephrotic syndrome. Further topics were hematurias, membranoproliferative glomerulonephritis [2], vesicoureteral reflux [33, 39], other nephritides [29], and proteinurias.

Table II. ESPN members involved in interventional cooperative trials (from 101 answers)

Topic	Number of members
MCNS	27
Vesicoureteral reflux	11
Hemolytic-uremic syndrome	10
Focal segmental glomerulosclerosis	3
Other glomerulonephritides	3
Pyelonephritis	1
Total	55

About the same number of ESPN members (55) participated actively in interventional therapeutic trials (table II). The most extensive disease studied with this method was minimal change nephrotic syndrome (MCNS). Other studies were concerned with vesicoureteral reflux, hemolytic uremic syndrome, focal segmental glomerulosclerosis, other glomerulonephritides, and pyelonephritis.

Nine therapeutic trials are ongoing by ESPN members, and their topics are listed in table III. Most of them are concerned with minimal change nephrotic syndrome in which different types of treatment are compared. The results have appeared in the literature [3–6, 16, 23]. Other interventional studies are concerned with the treatment of hemolytic uremic syndrome [30, 37] and of vesicoureteral reflux [13, 14, 24–27].

As a final point of the questionnaire some comments concerning cooperative multicenter studies were asked for. The replies show that the great majority of those who commented were in favor of multicenter studies, and that only 5% were against such studies. The comments in favor regarded the results of such studies as important, and stated that they 'are essential', 'costly but in the long run effective', 'very useful, sometimes compulsory, always very, very, difficult', and 'are badly needed and require the work of very devoted people'.

Beneficial effects of multicenter studies on the participating centers were seen by other comments, which stated that they are 'useful in order to improve the standard of patient's care', and that they represent 'one method for the participation of small units in clinical studies'. A further beneficial side effect of multicenter studies is the standardization of terms

Table III. Recent or ongoing prospective, controlled interventional multicenter studies of ESPN members

Topic	Group
Levamisole in the treatment of steroid-dependent MCNS	BAPN
Long prednisone (12 weeks) vs. standard (8 weeks) in the initial MCNS	APN
Short steroid (2.5 months) vs. standard (4.5 months) in the first stage of MCNS	CFNP
Ciclosporin in cortico-dependent and resistant NS	SINP
Ciclosporin in steroid-resistant MCNS and FSGS	APN
HUS treatment with plasma vs. symptomatic treatment	CFNP
HUS treatment with plasma vs. normal saline	SINP
Medical vs. surgical treatment of VUR	IRSC
Operative vs. nonoperative treatment of VUR	BRSG

BAPN = British Association of Paediatric Nephrology; APN = Arbeitsgemeinschaft für Pädiatrische Nephrologie (FRG); CFNP = Club français de néphrologie pédiatrique; SINP = Società Italiana di Nephrologia Paediatrica; IRSC = International Reflux Study in Children; BRSG = Birmingham Reflux Study Group.

and regimens [15], which originally was the consequence of the successful multicenter study of the International Study of Kidney Disease in Children (ISKDC) [9, 10].

More critical comments were concerned with the efficiency of such studies in relation to the great efforts needed to participate. It was stated that 'most trials are too complex and time consuming', and 'that it is a real danger, to be too ambitious'. Furthermore, one should secure that the 'basic economic guidelines be known before beginning', and 'if the studies become too large, they lose efficiency'.

A few comments were very sceptical and expressed the opinion that multicenter studies 'generally are never satisfactorily completed, if more than 1–2 centers are involved'. Furthermore, it is said that 'if a lot of groups are involved, it is very difficult to give individual groups sufficient incentive'. Another somewhat sarcastic comment stated that 'international trials are never productive'.

The last opinion is counteracted by two other comments which were in favor of multicenter and international studies. One stated that studies should 'be encouraged and (today) lacked in ESPN', and the other one wished 'to see ESPN studies' started. I personally would agree with the last statement and could indeed envision some multicenter studies in which all ESPN members could participate.

Essentials for Controlled Multicenter Studies

From our own experience I would like to outline some essential aspects which need to be fulfilled in order to set up and run a successful multicenter study. There are three stages in the establishment and conduct of a controlled trial. Each stage is equally important and has to be thoroughly fulfilled otherwise such an endeavor will be doomed to fail. These stages are: (1) careful planning; (2) continuous surveillance, and (3) clear endpoint with evaluation of results.

Planning

The most crucial phase of a multicenter study is the planning [34]. This needs, in a first step, a core of highly qualified people, who are experts on the field to be studied and are well acquainted with the relevant scientific literature. They should define the aim of the study, the identification of the patients in the experimental and control groups, in order to achieve a high degree of homogeneity of the study population, and the procedure of random allocation [12, 21]. Furthermore, they should describe the interventional regimen precisely, the control procedures during the study and the methods for biostatistical evaluation [36]. Finally, they should discuss the economical basis [18] and the ethical aspects [20, 32, 38] of the proposed study, including requirements for informed consent and ethical committees.

The aim of a multicenter study should be as simple as possible. One has to realize that the more complex the questions of a study, the more complicated and difficult it will be to reach a definite conclusion [19]. Therefore, in order to get a clear-cut answer, just ask for a simple question, which may be, for instance, the occurrence of a relapse in MCNS, or the entry into renal death or defined degree of renal insufficiency in glomerular disorders.

The necessity for a distinct characterization of the patient to be studied cannot be overemphasized. In a multicenter study one needs very stringent definitions of the patient and his disease state, otherwise one will be lost in an inhomogeneous and amorphous study population. For

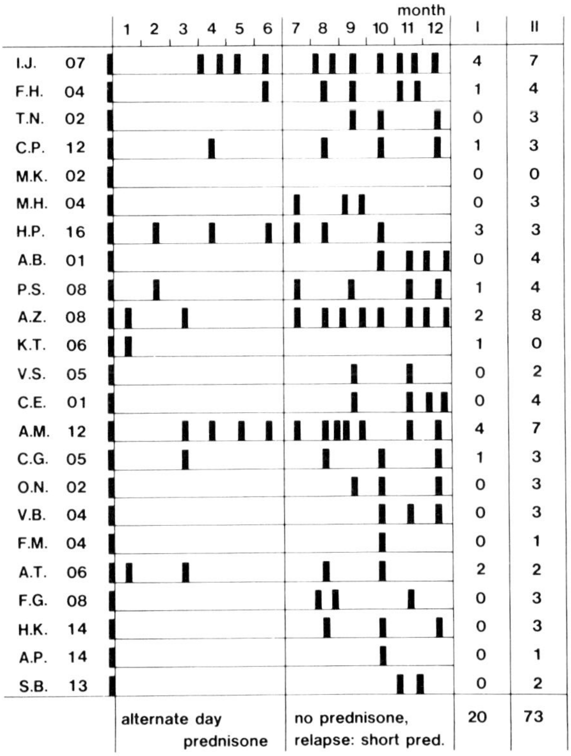

Fig. 1. Number of relapses (black bars) of children with frequently relapsing minimal change nephrotic syndrome under alternate-day or short prednisone treatment. Steroid Study I of APN [from ref. 4].

instance, when we in the Arbeitsgemeinschaft für Pädiatrische Nephrologie (APN) 1977 evaluated the results of our first Steroid Study, we found out that the group of patients we had studied was not as homogeneous as we thought [3, 4]. We had defined the study patients as 'frequent relapsers' of MCNS according to the definitions given by ISKDC [1, 28]. In those, we studied three different types of steroid regimens, i.e. the intermittent regimen, the alternate-day regimen, and the short regimen. Thus, we thought that we treated a rather homogeneous group of patients with different treatment regimens. However, as one can recognize from the original data shown in figure 1, there seemed to be at least two subgroups of patients within the group of frequent relapsers. One group under the short predni-

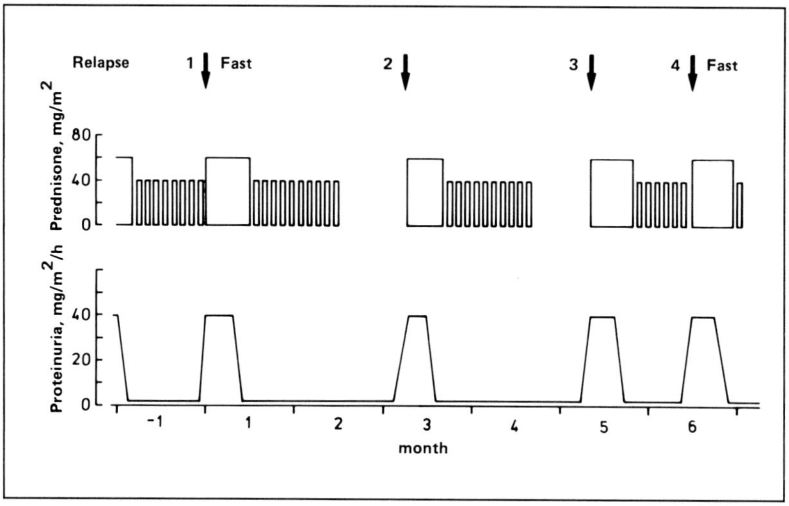

Fig. 2. Relapse rates of a child with nephrotic syndrome which qualify the patient as steroid dependent.

sone regimen had only 1–4 relapses per 6 months, the other group, however, had the double amount, i.e. 7–8 relapses. Therefore, the need to differentiate between these two subgroups became obvious. When we planned the next study, the Cytotoxic Drug Study I, we drew out the consequences and tried to characterize these two subgroups of frequent relapsers, in order to randomize them separately [5]. Thus, we defined the frequent relapser without steroid dependency as one who was characterized by 2 relapses within 6 months of the initial response, or 4 relapses within a 1-year period, without fulfilling the criteria of steroid dependency. We defined the steroid-dependent patient as one who had two consecutive relapses during the standard relapse prednisone therapy or within 14 days after cessation of steroid therapy (which we called 'fast relapse'), or who had 2 fast relapses out of 4 relapses in a period of 6 months.

The latter relapse rate is depicted graphically in figure 2. In the upper panel the prednisone treatment is depicted, and in the lower panel the proteinuria indicative of a relapse. The first relapse occurred during the alternate-day prednisone treatment of the last relapse, i.e. was a fast relapse, the second and third relapses were regular, the fourth relapse within the 6-month period again was a fast relapse. In this patient, there-

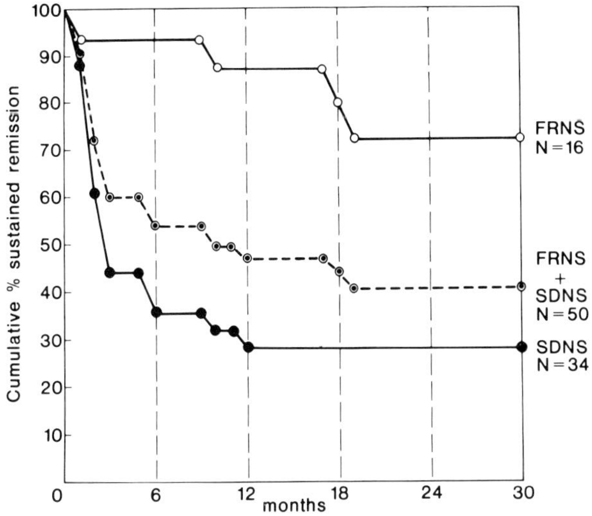

Fig. 3. The response of children with frequently relapsing nephrotic syndrome without steroid dependency (FRNS) and those with steroid-dependent nephrotic syndrome (SDNS) to an 8-week course of cyclophosphamide (2 mg/kg/day) or chlorambucil (0.15 mg/kg/day). Results of Cytotoxic Drug Study I of APN [5].

fore, the definition of steroid-dependent nephrotic syndrome was fulfilled.

These definitions, of course, are always somewhat arbitrary and often boring and cumbersome. But in multicenter studies and especially in a prospective trial one has to agree on definitions before the study begins and then keep to them strictly throughout the study. After the end of the study one should evaluate whether the definitions were correct or at least useful, or not. We have used the definitions of the frequent relapsers without steroid dependency and with steroid dependency prospectively in our trial, and found them very useful. In the first group the majority of patients experienced long-lasting remissions after an 8-week course of chlorambucil or cyclophosphamide, in the latter group the majority of patients relapsed again early after cessation of cytotoxic drug treatment (fig. 3). Because of this outcome we kept the definitions of frequent relapsers with and without steroid dependency, and would propose that other studies on similar topics should also use them.

There is another important point to be mentioned in this context. The definition of a patient and his disease is intimately connected with the type of treatment he receives. If one uses other regimens of steroid treatment than the standard treatment of the ISKDC or APN for the initial attack of a nephrotic syndrome or for the relapse, one cannot use the definitions of patients, given by these groups, because the different steroid treatment produces different study patients. Therefore, the definition of a patient has to include the type of treatment, which, of course, changes the natural history of the disease, Thus, if we would decide to start a therapeutic multicenter trial of ESPN on MCNS we first would have to agree on a common standard steroid treatment regimen.

When the first draft of a study protocol is completed, then in a second step, all potential participants should have the opportunity to discuss the protocol at length and in detail. This step seems to me as important as the first one, since by this, each participant will be directly involved in the decision making for the protocol. By this he will exactly understand the reason for the definitions and regimens, and identify himself with the content of the protocol. This is the best guarantee that the rules of the protocol will be kept throughout the study in each center, wich is a precondition for a sucessful trial.

You may realize that the planning of a protocol for a cooperative multicenter study is a burdensome endeavor. It is not easy on a national basis, but will even be more complicated if it will expand to international cooperation. One has to estimate that it will take at least 1 year before a study can be started once the definite decision to do so was made.

Surveillance

When the study has started, continuous surveillance is essential. This includes the control and feedback by a central office which checks continuously all incoming data and, even more important, asks for missing items. The participants should receive their fundings, if this is a part of the protocol. They should be informed regularly on the progress of the study. For this purpose they should be able to meet at least twice a year with the central office. They should have enough time to discuss the different aspects of the study and participate in the results of preliminary evaluations of the study. This is the best way to keep the interest in the study alive, and it encourages the participants to continue their efforts. One has always to realize that the burden of a cooperative study can sometimes be very heavy, and that for a long time there seems to be no reward for the single participant or the participating center. The regular meetings, therefore, are mandatory for a successful course of a multicenter study. A sequential analysis of the results is important in order not to miss a definite trend, which then would lead to the question whether the continuation of the study is still justified for ethical reasons. New decisions, then, have to be carried out by all participants.

Endpoint

The last essential stage in the conduct of a cooperative multicenter trial is the endpoint and the critical evaluation of the results. The endpoint of a study, of course, has to be defined already in the planning. However, earlier termination could become necessary for ethical reasons, if one group of patients do profit clearly more than the other from the treatment [31]. It should be asked, whether the necessary number of patients is reached, the study groups are comparable, unforeseen biases could be avoided, the interventional procedures are completely fulfilled, and the compliance of patients could be secured. The last point is very important, since it is well known that the longer a study is running the more questionable a compliance will be. If these essentials have been answered then the statistical evaluation should take place, and this in an appropriate way. In this context, I would like to quote a word from one of the statistician fathers of controlled clinical trials, Sir Austin Bradford Hill, who said in his Marc Daniels lecture in 1963, that 'in a controlled trial, there is no need in the search for precision to throw common sense out of the window' [22]. Therefore, the evaluation of data and the statistical calculations are always an interdisciplinary endeavor.

The last point concerns the timely publications of results. Publications of multicenter trials poses problems, as pointed out by Barnett [11]. Published as reports of the group, identifying individual participants only as members of the study diminishes the personal credit to each investigator, especially important to younger members. The professional reward for an investigator participating in a multicenter trial is less than if he were acting alone. It is therefore important to clarify authorship policy from the start in order to avoid unnecessary misunderstandings at a later stage. Furthermore, one has to realize that publications of multicenter studies are always delayed by the need to obtain a consensus of all participants.

In summary, clinical trials are powerful tools in the scientific evaluation of treatments or interventional procedures. They are necessary and widely accepted [40]. However, they are not the only way to come to a scientifically oriented conclusion, since they are very limited in their answers. In principle, a trial is only capable of providing a solution to those questions which it set out to answer initially. In this aspect a trial can be compared with a computer. A computer has a wonderful memory and excellent capacities to combine data, but it gives you only an answer for those items you have fed it before. A computer will never substitute the human brain, and the clinical trial will therefore never compensate for the individual observation and experience of a devoted physician with a critical mind. I think both approaches are necessary.

References

1 Abramowicz, M.; Arneil, G.C.; Barnett, H.L.; Barron, B.A.; Edelmann, C.M.; Gordillo-P., G.; Greifer, I.; Hallman, N.; Kobayashi, O.; Tiddens, H.A.: Controlled trial of azathioprine in children with nephrotic syndrome. Lancet *i:* 959–961 (1976).

2 Anders, D. for the Arbeitsgemeinschaft für Pädiatrische Nephrologie: Cooperative long term follow-up study of children with membranoproliferative glomerulonephritis (MPGN, 3 Types); in Murakami, Kitagawa, Yabuta, Sakai, Recent advances in pediatric nephrology, pp. 253–259 (Elsevier Science Publ. B.V., Excerpta Medica, Amsterdam-New York-Oxford 1987).

3 Arbeitsgemeinschaft für Pädiatrische Nephrologie: Alternate-day versus intermittent prednisone in frequently relapsing nephrotic syndrome. Lancet *i:* 401–403 (1979).

4 Arbeitsgemeinschaft für Pädiatrische Nephrologie: Alternate-day prednisone is more effective than intermittent prednisone in frequently relapsing nephrotic syndrome. Eur. J. Pediat. *135:* 229–237 (1981).

5 Arbeitsgemeinschaft für Pädiatrische Nephrologie: Effect of cytotoxic drugs in frequently relapsing nephrotic syndrome with and without steroid dependence. New Engl. J. Med. *306:* 451–454 (1982).

6 Arbeitsgemeinschaft für Pädiatrische Nephrologie – Oemar, B.S.; Brodehl, J.: Cyclophosphamide treatment of steroid dependent nephrotic syndrome: comparison of eight week with 12 week course. Archs Dis. Childh. *62:* 1102–1106 (1987).

7 Arbeitsgemeinschaft für Pädiatrische Nephrologie – Pistor, K.; Olbing, H.; Schärer, K.: Children with chronic renal failure in the Federal Republic of Germany. I. Epidemiology, modes of treatment, survival. Clin. Nephrol. *23:* 272–277 (1985).

8 Arbeitsgemeinschaft für Pädiatrische Nephrologie – Pistor, K.; Schärer, K.; Olbing, H.; Tamminen-Möbius, T.: Children with chronic renal failure in the Federal Republic of Germany. II. Primary renal diseases, age and intervals from early renal failure to renal death. Clin. Nephrol. *23:* 278–284 (1985).

9 Arneil, G.C.: An international cooperative study, societies of paediatric nephrology, and Henry Barnett. Pediat. Clins. N. Am. *18:* 349–354 (1971).

10 Barnett, H.L.: The natural and treatment history of glomerular diseases in children. What can we learn from International Cooperative Studies? A report of the Interna-

tional Study of Kidney Disease in Children. Proc. 6th Int. Congr. Nephrology, Florence 1975, pp. 470–485 (Karger, Basel 1976).
11 Barnett, H.L.: Philosophy and ethics of multicenter international controlled clinical trials in children; in Brodehl, Ehrich, Paediatric nephrology, pp. 30–35 (Springer, Berlin 1984).
12 Barnett, H.L.; Edelmann, C.M.: Is randomization always random? Pediatrics, Springfield *53:* 583 (1974).
13 Birmingham Reflux Study Group: A prospective trial of operative versus non operative treatment of severe vesico-ureteric reflux: two years' observation in 96 children; in Hodson, Heptinstall, Contributions to nephrology: reflux nephropathy update, vol. 39, pp. 169–185 (Karger, Basel 1984).
14 Birmingham Reflux Study Group: Prospective trial of operative versus non operative treatment of severe vesico-ureteric reflux two years' observation in 96 children. Br. med. J. *287:* 171–174 (1983).
15 Brodehl, J.: Nephrotic syndrome in children: diagnosis and treatment. Wld Pediat. Child Care *2:* 9–18 (1986).
16 Brodehl, J.; Krohn, H.-P.; Ehrich, J.H.H.: The treatment of minimal change nephrotic syndrome (lipoidnephrosis): cooperative studies of the Arbeitsgemeinschaft für Pädiatrische Nephrologie (APN). Klin. Pädiat. *194:* 162–165 (1982).
17 Champion, G.; Nivet, H.: L'insufficience rénale terminale chez les enfants de moins de 4 ans. Archs fr. Pédiat. *43:* 481–486 (1986).
18 Detsky, A.S.: Using economic analysis to determine the resource consequences of choices made in planning clinical trials. J. chron. Dis. *38:* 753–765 (1985).
19 Feinstein, A.R.: An additional basic science for clinical medicine. II. The limitations of randomized trials. Ann. intern. Med. *99:* 544–550 (1983).
20 Freedman, B.: Equipoise and the ethics of clinical research. New Engl. J. Med. *317:* 141–145 (1987).
21 Guyatt, G.; Sackett, D.; Wayne, D.; Taylor, D.W.; Chong, J.; Roberts, R.; Pugley, S.: Determining optimal therapy, randomized trials in individual patients. New Engl. J. Med. *34:* 889–892 (1986).
22 Hill, A.B.: Medical ethics and controlled trials. Br. med. J. *i:* 1043–1049 (1963).
23 Imbasciati, E.; Gusmano, R.; Edefonti, A.; Zucchelli, P.; Pozzi, C.; Grassi, C.; Della Volpe, M.; Perfumo, F.; Petrone, P.; Picca, M.; Claris Appiani, A.; Pasquali, S.; Ponticelli, C.: Controlled trial of methyl prednisolone pulses and low dose oral prednisone for the minimal change nephrotic syndrome. Br. med. J. *291:* 1305–1308 (1985).
24 International Reflux Study: Medical versus surgical treatment of VUR. A prospective international reflux study in children. Pediat. Radiol. *15:* 105–109 (1985).
25 International Reflux Study Committee: Medical versus surgical treatment of primary vesico-ureteral reflux. Pediatrics, Springfield *67:* 392–400 (1981).
26 International Reflux Study Committee: Medical versus surgical treatment of primary vesico-ureteral reflux. A prospective international reflux study in children. J. Urol. *125:* 277–283 (1981).
27 Internationale Reflux Studie: Vergleich von medikamentöser und chirurgischer Behandlung beim primären vesiko-uretero-renalen Reflux. Mschr. Kinderheilk. *129:* 316–323 (1981).

28 International Study of Kidney Disease in Children: Prospective, controlled trial of cyclophosphamide therapy in children with the nephrotic syndrome. Lancet *ii:* 423–427 (1974).

29 Leumann, E.P.; Briner, J.; Donckerwolke, R.A.M.; Kuijten, R.; Largiader, F.: Recurrence of focal segmental glomerulosclerosis in the transplanted kidney. Nephron *25:* 65–71 (1980).

30 Loirat, C.; Beaufils, F.; Sonsino, E.; Schlegel, N.; Guesnu, M.; Pillion, G.; Andre, J.L.; Broyer, M.; Guyot, C.; Habib, R.; Mathieu, H.: Traitement du syndrome hémolytique et urémique de l'enfant par l'urokinase. Archs fr. Pédiat. *41:* 15–19 (1984).

31 Meier, P.: Terminating a trial. The ethical problem. Clin. Pharmacol. Ther. *25:* 633–640 (1979).

32 Mongrovius, R.; Seyberth, H.W.; Medical, ethical and legal aspects of clinical trials in pediatrics. Eur. J. Pharmacol. *18:* 121–127 (1980).

33 Olbing, H.; Tamminen-Möbius, T. – for the International Reflux Study in Children: Stability of nonobstructive grade III or IV VUR correlates with frequency and type of renal scarring; in Murakami, Kitagawa, Yabuta, Sakai, Recent advances in pediatric nephrology (Excerpta Medica, Amsterdam 1987).

34 Peto, R.; Pike, M.C.; Armitage, P.; Breslow, N.E.; Cox, D.R.; Howard, S.V.; Mantel, N.; McPherson, K.; Peto, J.; Smith, P.G.: Design and analysis of randomized clinical trials requiring prolonged observation of each patient. I. Introduction and design. Br. J. Cancer *34:* 585–612 (1976).

35 Pistor, K.; Bachmann, H.-J.; Bläker, F.; Bulla, M.; Dippell, J.; Hake, H.; Holtvoeth, W.; Kirsch, W.; Lennert, T.; Müller-Wiefel, D.E.; Offner, G.; Olbing, H.; Schüler, H.-W.; Stehr, K.; Straub, E.; Tewes, I.; Arbeitsgemeinschaft für Pädiatrische Nephrologie: Akutes Nierenversagen im Kindesalter. Mschr. Kinderheilk. *126:* 328–330 (1978).

36 Pocock, S.J.; Hughes, M.D.; Lee, R.J.: Statistical problems in the reporting of clinical trials. A survey of three medical journals. New Engl. J. Med. *317:* 426–432 (1987).

37 Rizzoni, G.; Claris Appiani, A.; Edefonti, A.; Facchin, P.; Franchini, F.; Gusmano R.; Imbasciati, E.; Pavanello, L.; Perfumo, F.; Remuzzi, G.: Plasma infusion for hemolytic uremic syndrome in children: results of a multicenter controlled trial. J. Pediat. *112:* 284–290 (1988).

38 Schafer, A.: The ethics of the randomized clinical trial. New Engl. J. Med. *307:* 719–724 (1982).

39 Smellie, J.M.; Ransley, P.G.; Normand, I.C.S.; Prescod, N.; Edwards, D.: Development of new renal scars. A collaborative study. Br. med. J. *290:* 1957–1960 (1985).

40 Spodick, D.H.: The randomized controlled clinical trial. Am. J. Med. *73:* 420–425 (1982).

Johannes Brodehl, MD, Department of Paediatric Nephrology and Metabolic Disorders, Kinderklinik, Medizinische Hochschule Hannover, D-3000 Hannover 61 (FRG)

Contr. Nephrol., vol. 67, pp. 219–225 (Karger, Basel 1988)

Outcome of Isolated Haematuria in Children

S. Túri[a], *M. Visy*[b], *A. Vissy*[c], *V. Jászai*[d], *Zs. Czirbesz*[e], *I. Haszon*[a], *Zs. Szelid*[f], *I. Ferkis*[g]

[a] Department of Paediatrics, University Medical School, Szeged;
[b] Second Department of Paediatrics, University Medical School, Budapest;
[c] County Children's Hospital, Miskolc;
[d] Department of Paediatrics, University Medical School, Pécs;
[e] Heim Pál Children's Hospital, Budapest;
[f] County Children's Hospital, Györ;
[g] Buda Children's Hospital, Budapest, Hungary

Persistent and recurrent isolated haematuria is not uncommon in children. In the event of this single symptom, the question that arises is: What is the likelihood of a poor outcome, and therefore what intensive examinations should the children undergo, including invasive procedures such as radiological, isotope investigations and kidney biopsy? There are literature data which refer to a benign prognosis of isolated haematuria, but the number of patients was rather low in these studies, and their follow-up period was short [1–6]. Other reports found isolated haematuria to be potentially a progressive disease in 25% of the cases [7], but in the latter studies other signs and symptoms besides haematuria were not strictly excluded.

We report here a multicentre study of children with isolated haematuria. The aim of this study was to determine the long-term outcome and the proportion of patients in whom invasive diagnostic methods can be necessary.

Materials and Methods

Three hundred and fifty children presenting isolated haematuria at onset and persisting or recurring for at least 6 months were followed for a period of 2–17 years. These children were under the control of three university paediatric departments and 60 other hospitals. During this period the total number of nephrology patients was 11,600 in these paediatric clinics. Thus, these 350 cases with isolated haematuria comprised 3.02% of the overall group of children with kidney or urinary tract diseases. Those cases who presented other accompanying nephrological symptoms within 6 months following the first observation of haematuria were excluded from the study.

All patients with isolated haematuria were subjected to a nephrological examination and periodic controls at the paediatric nephrological clinics. During this, the following definitions were used:

(1) Haematuria: 20 or more RBC/0.9 mm^3 in a counting chamber in a fresh, uncentrifuged midstream specimen. (2) Proteinuria: more than 25 mg/dl. Quantitative determinations were made by the biuret method. (3) Urinary tract infection: bacterial growth in a bladder puncture urine sample, or at least 100,000 colonies/ml for any individual organism in two consecutive clean-voided midstream specimens.

Renal function examinations including endogenous creatinine clearance, quantitative protein and calcium excretion, and urinary concentrating ability by the DDAVP test [8] and blood pressure control were carried out in all cases. Kidney ultrasound and i.v. urography were performed in all patients, and ^{125}I or ^{131}I renography and micturitional cystography in suspected cases of obstructive uropathy.

Following an average time of 12 months, kidney biopsy was carried out in 47 cases because of the associated symptoms of glomerulonephritis or nephrotic syndrome. Audiograms were initially recorded in the cases suspected of having Alport's syndrome, and more recently for all patients. This was also carried out on family members of patients with Alport's syndrome, together with routine urinalysis and blood pressure control.

Results

At the time of the first observation of haematuria, the age distribution of patients was as follows: 2 years 8.6%, 3 years 13.9%, 4 years 16.5%, 5 years 16.5%, 6 years 11.3%, 7 years 7.8%, 8 years 8.6%, 9 years 4.6%, 10 years 6.5%, 11 years 4.6%, 12 years 1.9%, of all cases. The proportions of boys and girls were 49.3 and 50.6%, respectively. The distribution of patients according to the follow-up periods was the following: 2–5 years control 207 cases (59.1%), 5–10 years 120 (34%), 10–15 years 21 (6%), and 15 years 2 cases (0.57%). The type of haematuria in the first 6 months following presentation and the outcome are presented in table I. 83.4% of all the patients became symptomless or the haematuria remained isolated. The rest of the cases developed proteinuria mostly associated with azotaemia, representing a more serious glomerulopathy. Three of 7 patients with a decreased creatinine clearance ($C_{creat} < 50$ ml/min/1.73 m^2) had Alport's syndrome, 3 other children had therapy-resistant nephrotic syndrome, and in 1 case a haemangioblastoma developed. This child and another patient with Alport's syndrome died because of a serious intercur-

Table I. Type of haematuria at onset and the outcome in 350 patients with isolated haematuria

	Number of cases	%
Type of haematuria		
Microscopic	109	31.1
Macroscopic	98	28.0
Micro/macroscopic	143	40.8
Persistent	141	40.2
Recurrent	209	59.7
Outcome of haematuria		
Persistently symptomless (> 2 years)	163	46.6
Remained isolated, haematuria	129	36.8
Haematuria + proteinuria	53	16.6
Haematuria + hypercalciuria	68	19.4
Urolithiasis	49	14.0
Haematuria + hypertension + proteinuria	4	1.1
Tumour	2	0.57
Azotaemia		
C_{crt} < 10 ml/min/1.73 m^2	1	0.3
10–50 ml/min/1.73 m^2	6	1.7
50–75 ml/min/1.73 m^2	29	8.2
Died	2	0.57

rent infection. The occurrence of kidney tumour or hypertension was not very high.

Hypercalciuria or urolithiasis were diagnosed in the late stage of haematuria. None of these cases showed any clinical symptoms or signs of urolithiasis on X-ray film or isotope renography in the first 6 months of observation (table I).

In 56 children the clinical symptoms developed into those of a more serious glomerulopathy. In most of them (53/56), the haematuria was associated with proteinuria. Two patients had post-streptococcal glomerulonephritis, and in 7 cases the persistent isolated haematuria manifested as part of the Henoch-Schönlein syndrome. The renal biopsy materials from 47 patients were subjected to light microscopy, immunofluorescence and electron microscopy. Seventeen children had mesangial proliferative glomerulonephritis (MSGN), with different degrees of histological damage. Ten

Table II. Comparison of the outcome of disease and pattern of early haematuria in 27 patients

Haematuria	n	Azotaemia	Died	Symptom-free
Persistent	13	5	2	1
Recurrent	14	–	–	6
Microscopic	8	2	1	2
Macroscopic	3	–	–	1
Mixed-type	16	3	1	4

patients were diagnosed as having Alport's syndrome on the basis of combined pathological (biopsies studied by electron microscopy) and clinical criteria. Four further children exhibited basement membrane changes typical or suggestive of Alport's syndrome, and haematuria was found among their first-degree relatives, but these patients and their family members did not display a sensorineural hearing loss or other suggestive features. The further distribution of the histological diagnoses were as follows: focal proliferative glomerulonephritis (gn) 5, membranoproliferative gn 3, membranous gn 2, IgA gn 3, minimal change nephrotic syndrome 2, focal segmental glomerular sclerosis (FSGS) 1.

Sixteen children with a picture of chronic glomerulonephritis or subsequently developed nephrotic syndrome were treated with immunosuppressive therapy. Histologically, these cases corresponded to MSGN, membranoproliferative glomerulonephritis, membranous glomerulonephritis, FSGS or minimal change nephrotic syndrome. In 9 children a long-term remission was achieved, but in 1 patient with FSGS end-stage renal failure developed and the regular haemodialysis programme was started subsequently. In 6 further patients, the proteinuria and haematuria did not change. Two of them displayed a markedly decreased (10–50 ml/min/1.73 m^2), and 1 a moderately decreased (50–75 ml/min/1.73 m^2) creatinine clearance.

Apart from the children with renal failure, growth was normal and the blood pressure for age was normally distributed except for those noted as being hypertensive.

In one of the university paediatric departments which serves as a haemodialysis-transplantation centre, the pattern of early haematuria and the final outcome were compared in all 27 patients who were under control there from the beginning of haematuria or who were admitted because of a progressive renal disease, but these latter cases also started with persistent/recurrent isolated haematuria. Table II shows that persistent haematuria, particularly with a microscopic or mixed pattern, has a worse prognosis than recurrent haematuria.

Five of 7 children with severe azotaemia (creatinine clearance < 50 ml/min/1.73 m^2) had a persistent microscopic or persistent mixed-type haematuria at onset. In 4, this was associated with proteinuria (> 0.5 g/day) and 1 patient had hypertension later on ($> 150/100$ mm Hg). One patient with recurrent macroscopic haematuria had a haemangioblastoma in his solitary kidney. Another child with macroscopic haematuria improved later.

Discussion

The proportion of our patients with persistent or recurrent, isolated haematuria among the total nephrological cases is not very high, but, considering the unknown underlying disease and the final outcome, the attention that must be paid to these children is not insignificant. Although we found a low morbidity rate in our series, the final prospect for these patients is less favourable than that for cases found in a haematuria screening, particularly in view of the high percentages of azotaemia (10.2%) and severe glomerulopathies (15.8%). It is very likely that the possibility of morbidity and mortality will increase in certain adolescent boys with Alport's syndrome or in the therapy-resistant glomerulopathies of immunological origin.

The population served by our 16 hospitals and 3 university paediatric departments is approximately 8 million. In this respect our series is comparable with the study of 100 patients reported by Miller et al. [10]. The explanation of our larger number of cases may be the different systems of child health care in the two countries concerned. In the Leeds-Manchester study, two large paediatric centres were involved with only very selected patient material (those children who were under the control of general practitioners or other physicians were omitted). In Hungary the patients with persistent, recurrent haematuria are almost exclusively under hospital management, and therefore the number of our cases probably more closely reflects the true proportion of these patients in the overall population (43.75/1 million). Among our patients the mortality rate was 0.57%, and the incidence of hypertension requiring treatment was also lower (1.1%) than among the patients of Miller et al. [10]. Nevertheless, we had 1 patient with end-stage kidney failure and 4 other children with an appreciably decreased (10–50 ml/min/1.73 m^2) and a further 29 with a moderately decreased (50–75 ml/min/1.73 m^2) creatinine clearance. The final fate of these children and those with various kinds of glomerulopathy will only be known after a longer period. At any event, the occurrence of associated proteinuria increases progressively with time (14.5% in the first 5 years, 38.1% after 10 years).

In a high percentage of the children (14.0%), the haematuria proved to originate from urolithiasis. As most of these urolithiasis diagnoses were made in the 2nd to 5th years, it is very likely that these stones were present in a non-visualizable size at the time of admission, and grew slowly in the following years. These stone-forming cases are partly involved in the larger group of patients with hypercalciuria (19.4%), which may also be responsible for isolated haematuria. Nevertheless, none of these children developed hypertension or a marked decrease in creatinine clearance. We observed that most patients with relatively severe azotaemia had persistent microscopic haematuria (5/7) with or without intermittent macroscopic haematuria at presentation.

In the cases in which proteinuria accompanied the haematuria, histological examination showed an advanced glomerular lesion. Hypertension was a very rare complication of isolated haematuria, and kidney tumour occurred extremely unfrequently among the underlying diseases.

In summary, the prognosis of persistent/recurrent, isolated haematuria is fairly good, but the final outcome in most cases may be predictable only after a long-term follow-up. In some patients, urolithiasis and/or hypercalciuria can be discovered several years following the appearance of haematuria. Persistent microhaematuria in the beginning, with or without macrohaematuria episodes, and later on associated proteinuria and/or hypertension suggest a worse prognosis and these complications indicate the need for histological examinations. In most cases, intermittent macroscopic haematuria rarely leads to chronic renal failure. In suspected cases of Alport's syndrome, the examination of family members may be helpful, including audiometry, but an accurate diagnosis can be reached only by electron microscopy. The most obvious way to reduce the number of slowly deteriorating and unexpectedly uraemic cases is early hospital examination and long-term follow-up with appropriate treatment as soon as possible.

Acknowledgements

The authors wish to acknowledge the tremendous help of their colleagues Judit Kisbán, Judit Fehér, Gabriella Halász, Mihály Marcell, Erzsébet Kolman, Ilona Kiss, Erika Bujdosó, Mária Frank, Péter Sipos, István Marosváry, Ferenc Kádár, Júlia Kecskés, Ilona Jaksits, Agnes Miskolczi, and Erzsébet Ladányi from the collaborating hospitals who kindly contributed to this study.

References

1 Arneil, G.C.; Lam, C.N.; McDonald, M.: Recurrent haematuria in 17 children. Br. med. J. *ii:* 233 (1969).
2 Ayoub, E.M.; Verbier, R.L.: Benign recurrent haematuria. Am. J. Dis. Child. *109:* 217–223 (1965).
3 Bodian, M.; Black, J.A.; Kobayashi, N.; Lake, B.D.; Schuler, S.E.: Recurrent haematuria in childhood. Q. Jl Med. *34:* 359–382 (1965).
4 Glasgow, E.F.; Moncrieff, M.W.; White, R.H.R.: Symptomless haematuria in childhood. Br. med. J. *ii:* 687–692 (1970).
5 Pardo, V.; Berion, M.G.; Levi, D.F.; Strauss, J.: Benign primary haematuria. Clinicopathologic study of 65 patients. Am. J. Med. *67:* 817–822 (1979).
6 Schoeneman, M.J.; Earon, Y.; Spitzer, A.; Greifer, I.: Idiopathic persistent microscopic haematuria in children. Prognostic features. N.Y. State J. Med. *79:* 1714–1718 (1979).
7 Trachtman, H.; Weiss, R.A.; Bennett, B.; Greifer, I.: Isolated haematuria in children: indications for a renal biopsy. Kidney int. *25:* 94–99 (1984).
8 Aronson, A.S.; Svenningsen, N.W.: DDAVP test for estimation of renal concentrating capacity in infants and children. Archs Dis. Childh. *49:* 654–659 (1974).
9 Vehaskari, V.M.; Rapola, J.; Koskimies, O.; Savilahti, E.; Vilska, J.; Hallman, N.: Microscopic haematuria in schoolchildren. Epidemiology and clinicopathologic evaluation. J. Pediat. *95:* 676–684 (1979).
10 Miller, P.F.W.; Speirs, N.J.; Aparicio, S.R.; Lendon, M.; Savage, J.M.; Postlethwaite, R.J.; Brocklebank, J.T.; Houston, J.B.; Meadow, S.R.: Long term prognosis of recurrent haematuria. Archs Dis. Childh. *60:* 420–425 (1985).

S. Túri, MD, Department of Paediatrics, University Medical School,
Korányi str. 14-15, H–6725 Szeged (Hungary)

Contr. Nephrol., vol. 67, pp. 226–228 (Karger, Basel 1988)

Steroid-Sensitive Minimal Change Nephrotic Syndrome

Long-Term Follow-Up

M.A. Lewis, E.M. Baildom, N. Davies, I.B. Houston, R.J. Postlethwaite

The Royal Manchester Children's Hospital, Pendlebury, UK

Nephrotic syndrome is one of the most common nephrological disorders of childhood. Over 75% of the cases are associated with minimal change histology. The vast majority of these are steroid sensitive. We report the long-term follow-up of a group of patients who are both steroid sensitive and who have minimal change histology to provide an accurate prognosis for the group which account for the majority of children with nephrosis.

Patients and Methods

All the 68 patients with nephrotic syndrome referred to the Royal Manchester Children's Hospital between 1963 and 1976 who were both steroid responsive and who had minimal change histology were included. This permitted a minimum 10-year follow-up. Relevant information was taken from clinical records.

Analysis. The Student's t test, chi-square test and the correlation coefficients between variables were used as appropriate.

Results

Of the 68 patients 5 were lost to follow-up.

Death. Two patients died, 1 from peritonitis and septicaemia in the first attack and one from sagittal sinus thrombosis in the first relapse. The mortality rate was thus 3%.

Follow-Up. For the other 61 patients the mean period of follow-up was 14 years with a range of 10–21 years.

Hypertension and Renal Function. At the end of the follow-up period no patient was hypertensive (BP $> 97^{th}$ percentile for age) and all had a serum creatinine < 100 μmol/l.

Age and Sex. Modal age of onset was 2 years, mean age of onset 5 years. Male to female sex ratio was 3.5:1.

Treatment. Seventeen patients received steroid therapy only for the period of follow-up (group S) and 44 patients received cytotoxic drugs in addition (group S + C).

Relapses. Relapses occurred in 59 patients (97%). The rate of relapse, both in terms of the number of patients relapsing in each year and in terms of the number of relapses per patient per year, fell rapidly over the first 3–4 years of follow-up but thereafter changed little.

Three variables were age of onset, sex, and treatment. No one of these variables was totally dependent upon the other.

Age Effect. There was an inverse correlation between age of onset and the total number of relapses over the first 10 years after diagnosis ($r = -0.51$, $p < 0.001$), age of onset accounting for about 25% of the variability in total number of relapses. Those with an age of onset before their fifth birthday had significantly more relapses than older patients in each year after diagnosis ($p < 0.001$).

Sex Effect. Female sex conferred a small advantage, with a smaller proportion of girls relapsing in each year than boys ($p < 0.02$).

Treatment Effect. The rapid fall in rate of relapse over the first few years after diagnosis was most pronounced in group S+C, the percentage of patients relapsing per year over the first 3 years being significantly higher than that of group S ($p < 0.001$). Over the last 7 years there was no significant difference in the relapse rate of the two groups.

Remissions. Five of 30 patients who had both a 5-year remission and a subsequent 5-year follow-up relapsed, giving a $> 17\%$ probability of relapse after a 5-year remission. The longest individual periods of remission followed by relapse were 11 and 14 years.

Adulthood. Relapses continued into adult life. Twelve of 45 patients over the age of 16 years and 5 of 26 patients over 20 years were relapsing. Age of onset did not influence the risk of relapse as an adult.

Discussion

The results suggest that the long-term outlook is excellent though the vast majority of patients relapse and these relapses can continue into adult life. Young age of onset is associated with an increased frequency of relapse but not an increased risk of relapse as an adult. Long remission is not equivalent to cure. Mortality is associated with the complications of relapse and the oedematous state. If these are to be avoided, regular urine testing must be continued indefinitely.

M.A. Lewis, MD, The Royal Manchester Children's Hospital, Pendlebury, Manchester M27 1HA (UK)

Contr. Nephrol., vol. 67, pp. 229–232 (Karger, Basel 1988)

Evaluation of Prednisolone Pulse Therapy in Steroid-Resistant Nephrotic Syndrome

A Multicenter Collaborative Study

T. Wyszynska[a], *J. Ksiazek*[a], *M. Uszycka-Karcz*[b],
A. Kobierska-Szczepanska[c], *Z. Morawska*[d], *W. Zoch-Zwierz*[e]

Departments of Paediatric Nephrology, Child Health Center, [a]Warsaw;
Medical Schools in [b]Gdansk, [c]Zabrze, [d]Wroclaw and [e]Bialystok, Poland

High intravenous doses of methylprednisolone (pulse therapy) are used in children with different forms of glomerulopathies [2, 8, 9], including minimal change nephrotic syndrome [4, 6, 7].

The aim of this study was to evaluate the efficacy and safety of intravenous prednisolone pulse therapy in children with idiopathic nephrotic syndrome resistant to conventional steroid treatment.

Material and Methods

In the 5 participating clinics, 77 children with nephrotic syndrome were included in the study between 30.04.84 and 30.04.86. The criteria for patients to be included were as follows: (1) age between 12 months and 12 years; (2) resistance to at least 8 weeks of prednisolone therapy (2–3 mg/kg body weight per day) from the beginning of their disease (initial nonresponders) or during subsequent relapse of the nephrotic syndrome (late nonresponders); (3) normal blood pressure and renal function; (4) minimal change nephrotic syndrome (MCNS) or mesangial proliferative glomerulonephritis (MSGN) on renal biopsy.

The definitions of nephrotic syndrome, steroid responsiveness, remission and relapse proposed by the International Study of Kidney Disease in Children (ISKDC) [1, 5] were accepted, while the diagnosis of MCNS was also made in accordance with the ISKDC [5] and that of MSGN in accordance with Glassock et al. [3].

Among the 77 children fulfilling the trial entrance criteria, 48 were males and 29 females; 32 were below 24 months of age and 10 older than 7 years; 13 were classified as initial and 64 as late nonresponders. The diagnosis of MCNS was established in 40 patients and that of MSGN in 37 patients.

Table I. Effects of prednisolone pulse therapy in steroid-resistant nephrotic syndrome

Type of disease	n	Remission		No change	
		n	%	n	%
MCNS	40	33	82	7	18
MSGN	37	26	70	11	30
Initial nonresponders	13	8	62	5	38
Late nonresponders	64	51	80	13	20

Prednisolone (20 mg/kg body weight, up to a maximum dose of 1,000 mg) was given via a continuous i.v. drop infusion. The drug was administered on 3 consecutive days and thereafter on each alternate day until the urine became protein free. In 3 children only 3 infusions were needed, while 25 were treated with 4–6 pulses, and 46 with 7–12 pulses. Children who responded were treated later with a single oral dose of prednisolone (1 mg/kg body weight) daily for 4 weeks, and thereafter with the same dose on each alternate day for the next 4 weeks. Subsequently, prednisolone was withdrawn by a stepwise reduction of the dose during 4 weeks. All patients attended the outpatients department every 4 weeks. The shortest period of follow-up after completion of the treatment was 9 months.

Results

Remission was obtained in 59 (76.6%) children, whereas 18 (23.4%) failed to respond. The responses are indicated in table I. Relapse of the nephrotic syndrome was observed in 32 children. The 'time to relapse' curve after cessation of the therapy is shown in figure 1. Long-lasting remission was achieved in 27 children: 19 remained in remission for more than 12 months, while 8 who did not complete the 1-year follow-up period were still in remission after at least 9 months. The children who failed to respond (7 with MCNS and 11 with MSGN) were subsequently treated with cytotoxic drugs: remission was obtained in 13, but 5 are still nephrotic. Mild and transient hypertension occurring during i.v. prednisolone administration (in 9 children) was the only side effect observed in this study.

In summary, 59 of 77 or 76.6% of children with steroid-resistant nephrotic syndrome and a pathology compatible with the diagnosis of

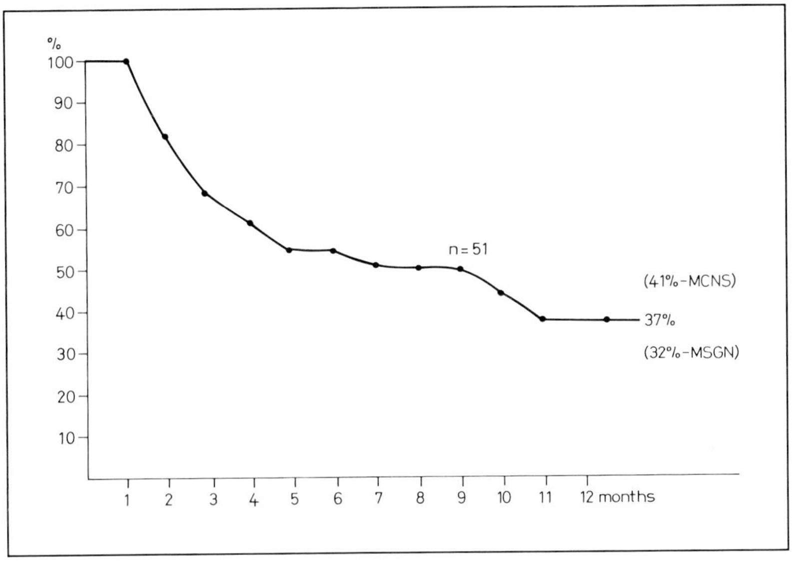

Fig. 1. Actuarial analysis of patients remaining in remission. 41% MCNS and 31% MSGN: percentages of children with minimal change nephrotic syndrome and mesangial proliferative glomerulonephritis, respectively, remaining in full remission 12 months after cessation of the therapy.

MCNS or MSGN responded to i.v. prednisolone pulse therapy. The response seems to be better in children with MCNS than in MSGN, and in late rather than in initial nonresponders. It can be concluded that this prednisolone pulse therapy was effective in the majority of these nephrotic children resistant to conventional steroid treatment. No side effects except mild and transient arterial hypertension were seen.

References

1 Brodehl, J.: Nephrotic syndrome in children: diagnosis and treatment. Wld Pediat. Child Care *2:* 9–18 (1986).

2 Cole, B.R.; Brocklebank, J.T.; Kienstra, R.A.; Kisaane, J.M.; Robson, M.A.: 'Pulse' methylprednisolone therapy in the severe glomerulonephritis. J. Pediat. *88:* 307–314 (1976).

3 Glassock, R.J.; Adler, S.G.; Ward, H.J.; Cohen, A.H.: Primary glomerular disease; in Brenner, The kidney, vol. II, pp. 929–1103 (Saunders, Philadelphia 1986).

4 Imbasciati, E.; Gusmano, R.; Edefonti, A.; Zucchelli, P.; Pozzi, C.; Grassi, C.; Della Volpe, M.; Perfumo, F.; Petrone, P.; Picca, M.; Claris Appiani, A.; Pasquali, P.; Ponticelli, C.: Controlled trial of methylprednisolone pulses and low dose oral prednisolone for the minimal change nephrotic syndrome. Br. med. J. *291:* 1305–1308 (1985).

5 International Study of Kidney Disease in Children: Primary nephrotic syndrome in children: clinical significance of histopathologic variants of minimal change and of diffuse mesangial hypercellularity. Kidney int. *20:* 765–771 (1985).

6 Krynski, J.; Wyszynska, T.: The effectiveness of intravenous prednisolone 'pulse' therapy in steroid-resistant and steroid-dependent nephrotic syndrome in children. Pediat. pol. *60:* 163–168 (1985).

7 Ponticelli, C.; Imbasciati, E.; Case, N.; Zucchelli, P.; Cagnoli, L.; Pasquali, S.: Intravenous methylprednisolone in minimal change nephrotic syndrome. Br. med. J. *280:* 685 (1980).

8 Robson, A.M.; Rose, G.M.; Cole, B.R.; Ingelfinger, J.R.: The treatment of severe glomerulopathies in children with intravenous methylprednisolone 'pulses'. Proc. 8th Int. Congr. Nephrol., Athens 1981, pp. 305–311 (Karger, Basel 1981).

9 Rose, G.M.; Cole, B.R.; Robson, A.M.: The treatment of severe glomerulopathies in children using high dose intravenous methylprednisolone pulses. Am. J. Kidney Dis. *1:* 148–156 (1981).

Prof. T. Wyszynska, Memorial Hospital, Child Health Center,
Al. Dzieci Polskich 20, 04–736 Warsaw (Poland)

Subject Index